Guide for the Care and Use of Laboratory Animals

Institute of Laboratory Animal Resources
Commission on Life Sciences
National Research Council

NATIONAL ACADEMY PRESS
Washington, D.C.   1996

NATIONAL ACADEMY PRESS • 2101 Constitution Avenue, NW • Washington, DC 20418

NOTICE: The project that is the subject of this report was approved by the Governing Board of the National Research Council, whose members are drawn from the councils of the National Academy of Sciences, National Academy of Engineering, and Institute of Medicine. The members of the committee responsible for the report were chosen for their special competences and with regard for appropriate balance.

This report has been reviewed by a group other than the authors according to procedures approved by a Report Review Committee consisting of members of the National Academy of Sciences, National Academy of Engineering, and Institute of Medicine.

This study was supported by the Comparative Medicine Program, National Center for Research Resources the Interagency Research Animal Committee, and the Office for Protection from Research Risks, National Institutes of Health/Department of Health and Human Services; the U.S. Department of Agriculture; and the Department of Veterans Affairs. The grant was awarded by the Comparative Medicine Program, National Center for Research Resources, and all agency funding was provided under grant NIH RR08779-02.

Core support is provided to the Institute of Laboratory Animal Resources by the Comparative Medicine Program, National Center for Research Resources, National Institutes of Health, through grant number 5P40RR0137; the National Science Foundation through grant number BIR-9024967; the U.S. Army Medical Research and Development Command, which serves as the lead agency for combined U.S. Department of Defense funding also received from the Human Systems Division of the U.S. Air Force Systems Command, Armed Forces Radiobiology Research Institute, Uniformed Services University of the Health Sciences, and U.S. Naval Medical Research and Development Command, through grant number DAMD17-93-J-3016; and by Research Project Grant #RC-1-34 from the American Cancer Society.

Any opinions, findings, and conclusions or recommendations expressed in this publication do not necessarily reflect the views of DHHS or other sponsors, nor does the mention of trade names, commercial products, or organizations imply endorsement by the U.S. government or other sponsors.

Note

The *Guide for the Care and Use of Laboratory Animals* was released to the sponsors and the public on January 2, 1996, in a prepublication form. After that, the Institute of Laboratory Animal Resources (ILAR) received comments from users and members of the Committee to Revise the *Guide*. The *Guide* has always been characterized as a living document, subject to modification with changing conditions and new information. That characterization results in a continuing emphasis on performance goals as opposed to engineering approaches. The use of performance goals places increasing responsibility on the user and results in greater enhancement of animal well-being; but performance goals require careful interpretation, whereas engineering goals leave no room for interpretation. With that difference in mind, the National Research Council and the appointed reviewers strove for accuracy and clarity. However, some errors and ambiguities were identified by readers of the prepublication copy. Some pointed out pagination, spelling, and reference errors. Others noted that some statements were being misinterpreted. After careful consideration, some changes have been made in this edition. For example, punctuation and spelling were corrected, and wording was changed to clarify meaning. An example of changes for clarification is replacement of the word "develop" with "review and approve" in descriptions of animal care and use committee (IACUC) oversight of housing plans, sanitation, and bedding selection; these are responsibilities of animal-care personnel, not of the IACUC, as the word "develop" might have implied. The discussion of monitoring of food and fluid restriction in small animals was clarified by addition of the phrase "such as rodents." Appendix B (Selected Organizations Related to Laboratory Animal Science) of the review copy that was sent to reviewers requested advice from reviewers regarding what organizations should be listed; some were added in the prepublication copy and others later. A footnote added to page 2 and referred to in three places reminds readers that the *Guide* is written for a broad international audience some of whom are not covered by either the Public Health Service Policy on Humane Care and Use of Laboratory Animals or the Animal Welfare Regulations but that those who are covered by these rules must abide by them even when the *Guide* recommends a different approach. That admonition is provided throughout the *Guide,* but its placement in the introduction was thought important. ILAR believes that each of these changes will help users to interpret and apply the recommendations as intended. *There was no substantial change in the content of the prepublication version.*

The National Academy of Sciences is a private, nonprofit, self-perpetuating society of distinguished scholars engaged in scientific and engineering research, dedicated to the furtherance of science and technology and to their use for the general welfare. Upon the authority of the charter granted to it by the Congress in 1863, the Academy has a mandate that requires it to advise the federal government on scientific and technical matters. Dr. Bruce Alberts is president of the National Academy of Sciences.

The National Academy of Engineering was established in 1964, under the charter of the National Academy of Sciences, as a parallel organization of outstanding engineers. It is autonomous in its administration and in the selection of its members, sharing with the National Academy of Sciences the responsibility for advising the federal government. The National Academy of Engineering also sponsors engineering programs aimed at meeting national needs, encourages education and research, and recognizes the superior achievements of engineers. Dr. Harold Liebowitz is president of the National Academy of Engineering.

The Institute of Medicine was established in 1970 by the National Academy of Sciences to secure the services of eminent members of appropriate professions in the examination of policy matters pertaining to the health of the public. The Institute acts under the responsibility given to the National Academy of Sciences by its congressional charter to be an adviser to the federal government and upon its own initiative to identify issues of medical care, research, and education. Dr. Kenneth I. Shine is president of the Institute of Medicine.

The National Research Council was established by the National Academy of Sciences in 1916 to associate the broad community of science and technology with the Academy's purposes of furthering knowledge and advising the federal government. Functioning in accordance with general policies determined by the Academy, the Council has become the principal operating agency of both the National Academy of Sciences and National Academy of Engineering in the conduct of their services to the government, the public, and the scientific and engineering communities. The Council is administered jointly by both Academies and the Institute of Medicine. Dr. Bruce Alberts and Dr. Harold Liebowitz are chairman and vice-chairman, respectively, of the National Research Council.

INSTITUTE OF LABORATORY ANIMAL RESOURCES COUNCIL

The Institute of Laboratory Animal Resources (ILAR) was founded in 1952 under the auspices of the National Research Council. A component of the Commission on Life Sciences, ILAR develops guidelines and disseminates information on the scientific, technological, and ethical use of animals and related biological resources in research, testing, and education. ILAR promotes high-quality, humane care of animals and the appropriate use of animals and alternatives. ILAR functions within the mission of the National Academy of Sciences as an advisor to the federal government, the biomedical research community, and the public.

COMMISSION ON LIFE SCIENCES

Preface

The *Guide for the Care and Use of Laboratory Animals* (the *Guide*) was first published in 1963 under the title *Guide for Laboratory Animal Facilities and Care* and was revised in 1965, 1968, 1972, 1978, and 1985. More than 400,000 copies have been distributed since it was first published, and it is widely accepted as a primary reference on animal care and use. The changes and new material in this seventh edition are in keeping with the belief that the *Guide* is subject to modification with changing conditions and new information.

The purpose of the *Guide*, as expressed in the charge to the Committee to Revise the *Guide for the Care and Use of Laboratory Animals*, is to assist institutions in caring for and using animals in ways judged to be scientifically, technically, and humanely appropriate. The *Guide* is also intended to assist investigators in fulfilling their obligation to plan and conduct animal experiments in accord with the highest scientific, humane, and ethical principles. The recommendations are based on published data, scientific principles, expert opinion, and experience with methods and practices that have proved to be consistent with high-quality, humane animal care and use.

Previous editions of the *Guide* were supported solely by the National Institutes of Health (NIH) and published by the Government Printing Office. As an indication of its wide use, this edition was financially supported by NIH, the Department of Agriculture, and the Department of Veterans Affairs and was published by the National Academy Press.

The *Guide* is organized into four chapters on the major components of an animal care and use program: institutional policies and responsibilities; animal environment, housing, and management; veterinary medical care; and physical plant. Responsibilities of institutional officials, institutional animal care and use committees, investigators, and veterinarians are discussed in each chapter.

In 1991, an ad hoc committee appointed by the Institute of Laboratory Animal Resources (ILAR) recommended that the *Guide* be revised. The Committee to Revise the *Guide for the Care and Use of Laboratory Animals* was appointed in 1993 by the National Research Council; its 15 members included research scientists, veterinarians, and nonscientists representing bioethics and the public's interest in animal welfare.

Before revision began, written and oral comments on the *Guide* were solicited widely from the scientific community and the general public. Open meetings were held in Washington, D.C., on December 1, 1993; in San Francisco, California, on February 2, 1994; and in St. Louis, Missouri, on February 4, 1994. Comments made at those meetings and written comments were considered by the committee and contributed substantially to this revision of the *Guide*.

The committee acknowledges the contributions of William I. Gay and Bennett J. Cohen in the development of the original *Guide*. In 1959, Animal Care Panel (ACP) President Cohen appointed the Committee on Ethical Considerations in the Care of Laboratory Animals to evaluate animal care and use. That committee was chaired by Dr. Gay, who soon recognized that the committee could not evaluate animal-care programs objectively without appropriate criteria on which to base its evaluations; that is, standards were needed. The ACP executive committee agreed, and the Professional Standards Committee was appointed. NIH later awarded the ACP a contract to "determine and establish a professional standard for laboratory animal care and facilities." Dr. Cohen chaired the ACP Animal Facilities Standards Committee, which prepared the first *Guide for Laboratory Animal Facilities and Care*.

The Committee to Revise the *Guide for the Care and Use of Laboratory Animals* expresses its appreciation to the Animal Welfare Information Center, National Agricultural Library, U.S. Department of Agriculture, for its assistance in compiling bibliographies and references. This task would have been quite formidable without their help. Appreciation is also extended to the reviewers of the volume, to Norman Grossblatt for editing the manuscript, to Carol Rozmiarek for providing exemplary secretarial assistance and preparing multiple drafts, and to Thomas L. Wolfle, who managed the process from beginning to end.

Readers who detect errors of omission or commission are invited to send corrections and suggestions to the Institute of Laboratory Animal Resources, National Research Council, 2101 Constitution Avenue, NW, Washington, DC 20418.

Derrell Clark, *Chairman*
Committee to Revise the *Guide for the Care and Use of Laboratory Animals*

Contents

Introduction

This edition of the *Guide for the Care and Use of Laboratory Animals* (the *Guide*) strongly affirms the conviction that all who care for or use animals in research, teaching, or testing must assume responsibility for their well-being. The *Guide* is applicable only after the decision is made to use animals in research, teaching, or testing. Decisions associated with the need to use animals are not within the purview of the *Guide*, but responsibility for animal well-being begins for the investigator with that decision. Additional responsibilities of the investigator, and other personnel, are elaborated in Chapter 1.

The goal of this *Guide* is to promote the humane care of animals used in biomedical and behavioral research, teaching, and testing; the basic objective is to provide information that will enhance animal well-being, the quality of biomedical research, and the advancement of biologic knowledge that is relevant to humans or animals. The use of animals as experimental subjects in the 20th century has contributed to many important advances in scientific and medical knowledge (Leader and Stark 1987). Although scientists have also developed nonanimal models for research, teaching, and testing (NRC 1977; see Appendix A, "Alternatives"), these models often cannot completely mimic the complex human or animal body, and continued progress in human and animal health and well-being requires the use of living animals. Nevertheless, efforts to develop and use scientifically valid alternatives, adjuncts, and refinements to animal research should continue.

In this *Guide*, laboratory animals include any vertebrate animal (e.g., traditional laboratory animals, farm animals, wildlife, and aquatic animals) used in research, teaching, or testing. When appropriate, exceptions or specific emphases for farm animals are provided. The *Guide* does not specifically address farm animals used in agricultural research or teaching, wildlife and aquatic animals studied in natural settings, or invertebrate animals used in research; however, many of the general principles in this *Guide* apply to these species and situations.

REGULATIONS, POLICIES, AND PRINCIPLES

This *Guide* endorses the responsibilities of investigators as stated in the *U.S. Government Principles for Utilization and Care of Vertebrate Animals Used in Testing, Research, and Training* (IRAC 1985; see Appendix D). Interpretation and application of those principles and this *Guide* require professional knowledge. In summary, the principles encourage

• Design and performance of procedures on the basis of relevance to human or animal health, advancement of knowledge, or the good of society.
• Use of appropriate species, quality, and number of animals.
• Avoidance or minimization of discomfort, distress, and pain in concert with sound science.
• Use of appropriate sedation, analgesia, or anesthesia.
• Establishment of experimental end points.
• Provision of appropriate animal husbandry directed and performed by qualified persons.
• Conduct of experimentation on living animals only by or under the close supervision of qualified and experienced persons.

In general, the principles stipulate responsibilities of investigators, whose activities regarding use of animals are subject to oversight by an institutional animal care and use committee (IACUC).

Animal facilities and programs should be operated in accord with this *Guide,* the Animal Welfare Regulations, or AWRs (CFR 1985); the Public Health Service Policy on Humane Care and Use of Laboratory Animals, or PHS Policy (PHS 1996); and other applicable federal (Appendixes C and D), state, and local laws, regulations, and policies.[1] Supplemental information on breeding, care, management, and use of selected laboratory animal species is available in other publications prepared by the Institute of Laboratory Animal Resources (ILAR) and other organizations (Appendix A). References in this *Guide* provide the

---

[1]Users are reminded that the *Guide* is written for a diverse group of national and international institutions and organizations, many of which are covered by neither the AWRs nor the PHS Policy. On a few matters, the *Guide* differs from the AWRs and the PHS Policy; users regulated by the AWRs or the PHS Policy must comply with them.

reader with additional information that supports statements made in the *Guide* or presents divergent opinions.

EVALUATION CRITERIA

The *Guide* charges users of research animals with the responsibility of achieving specified outcomes but leaves it up to them how to accomplish these goals. This "performance" approach is desirable because many variables (such as the species and previous history of the animals, facilities, expertise of the people, and research goals) often make prescriptive ("engineering") approaches impractical and unwarranted. Engineering standards are sometimes useful to establish a baseline, but they do not specify the goal or outcome (such as well-being, sanitation, or personnel safety) in terms of measurable criteria as do performance standards.

The engineering approach does not provide for interpretation or modification in the event that acceptable alternative methods are available or unusual circumstances arise. Performance standards define an outcome in detail and provide criteria for assessing that outcome, but do not limit the methods by which to achieve that outcome. This performance approach requires professional input and judgment to achieve outcome goals. Optimally, engineering and performance standards are balanced, thereby providing standards while allowing flexibility and judgment based on individual situations. Scientists, veterinarians, technicians, and others have extensive experience and information covering many of the topics discussed in this *Guide*. Research on laboratory animal management continues to generate scientific information that should be used in evaluating performance and engineering standards. For some issues, insufficient information is available, and continued research into improved methods of animal care and use is needed.

The *Guide* is deliberately written in general terms so that its recommendations can be applied in the diverse institutions and settings that produce or use animals for research, teaching, and testing; generalizations and broad recommendations are imperative in such a document. This approach requires that users, IACUCs, veterinarians, and producers use professional judgment in making specific decisions regarding animal care and use. Because this *Guide* is written in general terms, IACUCs have a key role in interpretation, oversight, and evaluation of institutional animal care and use programs. The question frequently arises as to how the words *must* and *should* are used in the *Guide* and how IACUCs should interpret their relative priority. In general, the verb *must* is used for broad programmatic or basic aspects that the Committee to Revise the *Guide* considers are imperative. The verb *should* is used as a strong recommendation for achieving a goal. However, the committee recognizes that individual circumstances might justify an alternative strategy.

FARM ANIMALS

Uses of farm animals in research, teaching, and testing are often separated into biomedical uses and agricultural uses because of government regulations (AWRs), institutional policies, administrative structure, funding sources, or user goals. That separation has led to a dual system with different criteria for evaluating protocols and standards of housing and care for animals of the same species on the basis of perceived biomedical or agricultural research objectives (Stricklin and Mench 1994). For some studies, this separation is clear. For example, animal models of human diseases, organ transplantation, and major surgery are considered biomedical uses; and studies on food and fiber production, such as feeding trials, are usually considered agricultural uses. However, the separation often is not clear, as in the case of some nutrition and disease studies. Administrators, regulators, and IACUCs often face a dilemma in deciding how to handle such studies (Stricklin and others 1990).

The use of farm animals in research should be subject to the same ethical considerations as the use of other animals in research, regardless of an investigator's research objectives or funding source (Stricklin and others 1990). However, differences in research goals lead to fundamental differences between biomedical and agricultural research. Agricultural research often necessitates that animals be managed according to contemporary farm-production practices for research goals to be reached (Stricklin and Mench 1994). For example, natural environmental conditions might be desirable for agricultural research, whereas control of environmental conditions to minimize variation might be desirable in biomedical research (Tillman 1994).

Housing systems for farm animals used in biomedical research might or might not differ from those in agricultural research. Animals used in either biomedical or agricultural research can be housed in cages or stalls or in paddocks or pastures (Tillman 1994). Some agricultural studies need uniform conditions to minimize environmental variability, and some biomedical studies are conducted in farm settings. Thus, the protocol, rather than the category of research, should determine the setting (farm or laboratory). Decisions on categorizing research uses of farm animals and defining standards for their care and use should be based on user goals, protocols, and concern for animal well-being and should be made by the IACUC. Regardless of the category of research, institutions are expected to provide oversight of all research animals and ensure that their pain and distress is minimized.

This *Guide* applies to farm animals used in biomedical research, including those maintained in typical farm settings. For such animals in a farm setting, the *Guide for the Care and Use of Agricultural Animals in Agricultural Research and Teaching* (1988), or revisions thereof, is a useful resource. Additional information regarding facilities and management of farm animals in an agricultural setting can be obtained from the Midwest Plan Service's *Structures and Environ-*

ment Handbook (1987) and from agricultural engineers or animal-science experts at state agricultural extension services and land-grant colleges and universities.

NONTRADITIONAL SPECIES

A species not commonly used in biomedical research is sometimes the animal model of choice because of its unique characteristics. For example, hibernation can be studied only in species that hibernate. An appropriate environment should be provided for nontraditional species, and for some species it might be necessary to approximate the natural habitat. Expert advice on the natural history and behavior of nontraditional species should be sought when such animals are to be introduced into a research environment. Because of the large number of nontraditional species and their varied requirements, this *Guide* cannot provide husbandry details appropriate to all such species. However, several scientific organizations have developed guides for particular species of nontraditional animals (e.g., ILAR and the Scientists Center for Animal Welfare, SCAW). A partial list of sources is available in Appendix A.

FIELD INVESTIGATIONS

Biomedical and behavioral investigations occasionally involve observation or use of vertebrate animals under field conditions. Although some of the recommendations listed in this volume are not applicable to field conditions, the basic principles of humane care and use apply to the use of animals living in natural conditions.

Investigators conducting field studies with animals should assure their IACUC that collection of specimens or invasive procedures will comply with state and federal regulations and this *Guide*. Zoonoses and occupational health and safety issues should be reviewed by the IACUC to ensure that field studies do not compromise the health and safety of other animals or persons working in the field. Guidelines for using animals in field studies prepared by professional societies are useful when they adhere to the humane principles of the *U.S. Government Principles for the Utilization and Care of Vertebrate Animals Used in Testing, Research, and Training* (Appendix D) and this *Guide* (see Appendix A, "Exotic, Wild, and Zoo Animals" and "Other Animals").

OVERVIEW

In an attempt to facilitate its usefulness and ease in locating specific topics, the organization of this edition of the *Guide* is slightly different from that of the preceding edition. Material from the preceding edition's Chapter 5, "Special Considerations," has been incorporated into Chapters 1-4. Genetics and nomenclature are now discussed in Chapter 2; facilities and procedures for animal research with hazardous agents and occupational health and safety are considered in Chapter 1. Recommendations for farm animals are incorporated throughout the text where appropriate.

This edition of the *Guide* is divided into four chapters and four appendixes. Chapter 1 focuses on institutional policies and responsibilities, including the monitoring of the care and use of animals, considerations for evaluation of some specific research procedures, veterinary care, personnel qualifications and training, and occupational health and safety; the latter section summarizes another National Research Council committee report (NRC In press) and includes information about facilities and procedures for animal research with hazardous agents. Chapter 2 focuses on the animals themselves and provides recommendations for housing and environment, behavioral management, husbandry, and population management, including discussions of identification, records, genetics, and nomenclature. Chapter 3 discusses veterinary medical care and responsibilities of the attending veterinarian; it includes recommendations relative to animal procurement and transportation, preventive medicine, surgery, pain and analgesia, and euthanasia. Chapter 4 discusses the physical plant, including functional areas and construction guidelines, with expanded discussions of heating, ventilation, and air-conditioning (HVAC) systems and facilities for aseptic surgery.

The appendixes in this edition remain largely the same as in the preceding edition. Appendix A contains an updated bibliography, categorized by topic; Appendix B lists selected organizations related to laboratory animal science; Appendix C presents federal laws relevant to animal care and use; and Appendix D provides the PHS endorsement of the *U.S. Government Principles for the Utilization and Care of Vertebrate Animals Used in Testing, Research, and Training* (IRAC 1985).

REFERENCES

CFR (Code of Federal Regulations). 1985. Title 9 (Animals and Animal Products), Subchapter A (Animal Welfare). Washington, D.C.: Office of the Federal Register.

Consortium for Developing a Guide for the Care and Use of Agricultural Animals in Agricultural Research and Teaching. 1988. Guide for the Care and Use of Agricultural Animals in Agricultural Research and Teaching. Champaign, Ill.: Consortium for Developing a Guide for the Care and Use of Agricultural Animals in Agricultural Research and Teaching.

IRAC (Interagency Research Animal Committee). 1985. U.S. Government Principles for Utilization and Care of Vertebrate Animals Used in Testing, Research, and Training. Federal Register, May 20, 1985. Washington, D.C.: Office of Science and Technology Policy.

Leader, R. W., and D. Stark. 1987. The importance of animals in biomedical research. Perspect. Biol. Med. 30(4):470-485.

Midwest Plan Service. 1987. Structures and Environment Handbook. 11th ed. rev. Ames: Midwest Plan Service, Iowa State University.

NRC (National Research Council). 1977. The Future of Animals, Cells, Models, and Systems in Research, Development, Education, and Testing. Proceedings of a Symposium of the Institute of Laboratory Animal Resources. Washington, D.C.: National Academy of Sciences. 341 pp.

NRC (National Research Council). In press. Occupational Health and Safety in the Care and Use of
Research Animals. A report of the Institute of Laboratory Animal Resources Committee on
Occupational Safety and Health in Research Animal Facilities. Washington, D.C.: National
Academy Press.

PHS (Public Health Service). 1996. Public Health Service Policy on Humane Care and Use of
Laboratory Animals. Washington, D.C.: U.S. Department of Health and Human Services, 28
pp. [PL 99-158, Health Research Extension Act, 1985]

Stricklin, W. R., and J. A. Mench. 1994. Oversight of the use of agricultural animals in university
teaching and research. ILAR News 36(1):9-14.

Stricklin, W. R., D. Purcell, and J. A. Mench. 1990. Farm animals in agricultural and biomedical
research in the well-being of agricultural animals in biomedical and agricultural research. Pp.
1-4 in Agricultural Animals in Research, Proceedings from a SCAW-sponsored conference,
September 6-7, 1990. Washington, D.C.: Scientist's Center for Animal Welfare.

Tillman, P. 1994. Integrating agricultural and biomedical research policies: Conflicts and opportunities. ILAR News 36(2):29-35.

1

Institutional Policies and Responsibilities

Proper care, use, and humane treatment of animals used in research, testing, and education (referred to in this *Guide* as animal care and use) require scientific and professional judgment based on knowledge of the needs of the animals and the special requirements of the research, testing, and educational programs. The guidelines in this section are intended to aid in developing institutional policies governing the care and use of animals.

Each institution should establish and provide resources for an animal care and use program that is managed in accord with this *Guide* and in compliance with applicable federal, state, and local laws and regulations, such as the federal Animal Welfare Regulations, or AWRs (CFR 1985), and Public Health Service Policy on Humane Care and Use of Laboratory Animals, or PHS Policy (PHS 1996). To implement the recommendations in this *Guide* effectively, an institutional animal care and use committee (IACUC) must be established to oversee and evaluate the program.

Responsibility for directing the program is generally given either to a veterinarian with training or experience in laboratory animal science and medicine or to another qualified professional. At least one veterinarian qualified through experience or training in laboratory animal science and medicine or in the species being used must be associated with the program. The institution is responsible for maintaining records of the activities of the IACUC and for conducting an occupational health and safety program.

MONITORING THE CARE AND USE OF ANIMALS

Institutional Animal Care and Use Committee

The responsible administrative official at each institution must appoint an IACUC, also referred to as "the committee," to oversee and evaluate the institution's animal program, procedures, and facilities to ensure that they are consistent with the recommendations in this *Guide*, the AWRs, and the PHS Policy. It is the institution's responsibility to provide suitable orientation, background materials, access to appropriate resources, and, if necessary, specific training to assist IACUC members in understanding and evaluating issues brought before the committee.

Committee membership should include the following:

• A doctor of veterinary medicine, who is certified (see American College of Laboratory Animal Medicine, ACLAM, Appendix B) or has training or experience in laboratory animal science and medicine or in the use of the species in question.

• At least one practicing scientist experienced in research involving animals.

• At least one public member to represent general community interests in the proper care and use of animals. Public members should not be laboratory-animal users, be affiliated with the institution, or be members of the immediate family of a person who is affiliated with the institution.

The size of the institution and the nature and extent of the research, testing, and educational programs will determine the number of members of the committee and their terms of appointment. Additional information about committee composition can be found in the PHS Policy and the AWRs.

The committee is responsible for oversight and evaluation of the animal care and use program and its components described in this *Guide*. Its functions include inspection of facilities; evaluation of programs and animal-activity areas; submission of reports to responsible institutional officials; review of proposed uses of animals in research, testing, or education (i.e., protocols); and establishment of a mechanism for receipt and review of concerns involving the care and use of animals at the institution.

The IACUC must meet as often as necessary to fulfill its responsibilities, but it should meet at least once every 6 months. Records of committee meetings and of results of deliberations should be maintained. The committee should review the animal-care program and inspect the animal facilities and activity areas at least once every 6 months. After review and inspection, a written report, signed by a majority of the IACUC, should be made to the responsible administrative officials of the institution on the status of the animal care and use program and other activities as stated herein and as required by federal, state, or local regulations and policies. Protocols should be reviewed in accord with the AWRs, the PHS Policy, *U.S. Government Principles for Utilization and Care of Vertebrate Animals Used in Testing, Research, and Training* (IRAC 1985; see Appendix D), and this *Guide* (see footnote, p. 2).

Animal Care and Use Protocols

The following topics should be considered in the preparation and review of animal care and use protocols:

- Rationale and purpose of the proposed use of animals.
- Justification of the species and number of animals requested. Whenever possible, the number of animals requested should be justified statistically.
- Availability or appropriateness of the use of less-invasive procedures, other species, isolated organ preparation, cell or tissue culture, or computer simulation (see Appendix A, "Alternatives").
- Adequacy of training and experience of personnel in the procedures used.
- Unusual housing and husbandry requirements.
- Appropriate sedation, analgesia, and anesthesia. (Scales of pain or invasiveness might aid in the preparation and review of protocols; see Appendix A, "Anesthesia, Pain and Surgery.")
- Unnecessary duplication of experiments.
- Conduct of multiple major operative procedures.
- Criteria and process for timely intervention, removal of animals from a study, or euthanasia if painful or stressful outcomes are anticipated.
- Postprocedure care.
- Method of euthanasia or disposition of animal.
- Safety of working environment for personnel.

Occasionally, protocols include procedures that have not been previously encountered or that have the potential to cause pain or distress that cannot be reliably controlled. Such procedures might include physical restraint, multiple major survival surgery, food or fluid restriction, use of adjuvants, use of death as an end point, use of noxious stimuli, skin or corneal irritancy testing, allowance of excessive tumor burden, intracardiac or orbital-sinus blood sampling, or the use of abnormal environmental conditions. Relevant objective information regarding the procedures and the purpose of the study should be sought from the literature, veterinarians, investigators, and others knowledgeable about the effects on animals. If little is known regarding a specific procedure, limited pilot studies designed to assess the effects of the procedure on the animals, conducted under IACUC oversight, might be appropriate. General guidelines for evaluation of some of those methods are provided in this section, but they might not apply in all instances.

Physical Restraint

Physical restraint is the use of manual or mechanical means to limit some or all of an animal's normal movement for the purpose of examination, collection of samples, drug administration, therapy, or experimental manipulation. Animals are restrained for brief periods, usually minutes, in most research applications.

Animals can be physically restrained briefly either manually or with restraint devices. Restraint devices should be suitable in size, design, and operation to minimize discomfort or injury to the animal. Many dogs, nonhuman primates (e.g., Reinhardt 1991, 1995), and other animals can be trained, through use of positive reinforcement, to present limbs or remain immobile for brief procedures.

Prolonged restraint, including chairing of nonhuman primates, should be avoided unless it is essential for achieving research objectives and is approved by the IACUC. Less-restrictive systems that do not limit an animal's ability to make normal postural adjustments, such as the tether system for nonhuman primates and stanchions for farm animals, should be used when compatible with protocol objectives (Bryant 1980; Byrd 1979; Grandin 1991; McNamee and others 1984; Morton and others 1987; Wakeley and others 1974). When restraint devices are used, they should be specifically designed to accomplish research goals that are impossible or impractical to accomplish by other means or to prevent injury to animals or personnel.

The following are important guidelines for restraint:

- Restraint devices are not to be considered normal methods of housing.
- Restraint devices should not be used simply as a convenience in handling or managing animals.
- The period of restraint should be the minimum required to accomplish the research objectives.
- Animals to be placed in restraint devices should be given training to adapt to the equipment and personnel.
- Provision should be made for observation of the animal at appropriate intervals, as determined by the IACUC.
- Veterinary care should be provided if lesions or illnesses associated with restraint are observed. The presence of lesions, illness, or severe behavioral change often necessitates temporary or permanent removal of the animal from restraint.

Multiple Major Surgical Procedures

Major surgery penetrates and exposes a body cavity or produces substantial impairment of physical or physiologic function. Multiple major survival surgical procedures on a single animal are discouraged but may be permitted if scientifically justified by the user and approved by the IACUC. For example, multiple major survival surgical procedures can be justified if they are related components of a research project, if they will conserve scarce animal resources (NRC 1990; see also footnote, p. 2), or if they are needed for clinical reasons. If multiple major survival surgery is approved, the IACUC should pay particular attention to animal well-being through continuing evaluation of outcomes. Cost savings alone is not an adequate reason for performing multiple major survival surgical procedures (AWRs).

Food or Fluid Restriction

When experimental situations require food or fluid restriction, at least minimal quantities of food and fluid should be available to provide for development of young animals and to maintain long-term well-being of all animals. Restriction for research purposes should be scientifically justified, and a program should be established to monitor physiologic or behavioral indexes, including criteria (such as weight loss or state of hydration) for temporary or permanent removal of an animal from the experimental protocol (Van Sluyters and Oberdorfer 1991). Restriction is typically measured as a percentage of the ad libitum or normal daily intake or as percentage change in an animal's body weight.

Precautions that should be used in cases of fluid restriction to avoid acute or chronic dehydration include daily recording of fluid intake and recording of body weight at least once a week (NIH 1990)—or more often, as might be needed for small animals, such as rodents. Special attention should be given to ensuring that animals consume a suitably balanced diet (NYAS 1988) because food consumption might decrease with fluid restriction. The least restriction that will achieve the scientific objective should be used. In the case of conditioned-response research protocols, use of a highly preferred food or fluid as positive reinforcement, instead of restriction, is recommended. Dietary control for husbandry or clinical purposes is addressed in Chapter 2.

VETERINARY CARE

Adequate veterinary care must be provided, including access to all animals for evaluation of their health and well-being. Institutional mission, programmatic goals, and size of the animal program will determine the need for full-time, part-time, or consultative veterinary services. Visits by a consulting or part-time veterinarian should be at intervals appropriate to programmatic needs. For specific responsibilities of the veterinarian, see Chapter 3.

Ethical, humane, and scientific considerations sometimes require the use of sedatives, analgesics, or anesthetics in animals (see Appendix A). An attending veterinarian (i.e., a veterinarian who has direct or delegated authority) should give research personnel advice that ensures that humane needs are met and are compatible with scientific requirements. The AWRs and PHS Policy require that the attending veterinarian have the authority to oversee the adequacy of other aspects of animal care and use. These can include animal husbandry and nutrition, sanitation practices, zoonosis control, and hazard containment.

PERSONNEL QUALIFICATIONS AND TRAINING

AWRs and PHS Policy require institutions to ensure that people caring for or using animals are qualified to do so. The number and qualifications of personnel required to conduct and support an animal care and use program depend on several factors, including the type and size of institution, the administrative structure for providing adequate animal care, the characteristics of the physical plant, the number and species of animals maintained, and the nature of the research, testing, and educational activities.

Personnel caring for animals should be appropriately trained (see Appendix A, "Technical and Professional Education"), and the institution should provide for formal or on-the-job training to facilitate effective implementation of the program and humane care and use of animals. According to the programmatic scope, personnel will be required with expertise in other disciplines, such as animal husbandry, administration, laboratory animal medicine and pathology, occupational health and safety, behavioral management, genetic management, and various other aspects of research support.

There are a number of options for the training of technicians. Many states have colleges with accredited programs in veterinary technology (AVMA 1995); most are 2-year programs that result in associate of science degrees, and some are 4-year programs that result in bachelor of science degrees. Nondegree training, with certification programs for laboratory animal technicians and technologists, can be obtained from the American Association for Laboratory Animal Science (AALAS). There are commercially available training materials that are appropriate for self-study (Appendix B). Personnel using or caring for animals should also participate regularly in continuing-education activities relevant to their responsibilities. They are encouraged to be involved in local and national meetings of AALAS and other relevant professional organizations. On-the-job training should be part of every technician's job and should be supplemented with institution-sponsored discussion and training programs and with reference materials applicable to their jobs and the species with which they work (Kreger 1995). Coordinators of institutional training programs can seek assistance from the Animal Welfare Information Center (AWIC) and ILAR (NRC 1991). The *Guide to the Care and Use of Experimental Animals* by the Canadian Council on Animal Care (CCAC 1993) and guidelines of some other countries are valuable additions to the libraries of laboratory animal scientists (Appendix B).

Investigators, technical personnel, trainees, and visiting investigators who perform animal anesthesia, surgery, or other experimental manipulations must be qualified through training or experience to accomplish these tasks in a humane and scientifically acceptable manner.

OCCUPATIONAL HEALTH AND SAFETY OF PERSONNEL

An occupational health and safety program must be part of the overall animal care and use program (CDC and NIH 1993; CFR 1984a,b,c; PHS Policy). The program must be consistent with federal, state, and local regulations and should focus on maintaining a safe and healthy workplace. The program will depend on the facility, research activities, hazards, and animal species involved. The National Research Council publication *Occupational Health and Safety in the Care and Use of Research Animals* (NRC In press) contains guidelines and references for establishing and maintaining an effective, comprehensive program (also see Appendix A). An effective program relies on strong administrative support and interactions among several institutional functions or activities, including the research program (as represented by the investigator), the animal care and use program (as represented by the veterinarian and the IACUC), the environmental health and safety program, occupational-health services, and administration (e.g., human resources, finance, and facility-maintenance personnel). Operational and day-to-day responsibility for safety in the workplace, however, resides with the laboratory or facility supervisor (e.g., principal investigator, facility director, or veterinarian) and depends on performance of safe work practices by all employees.

Hazard Identification and Risk Assessment

Professional staff who conduct and support research programs that involve hazardous biologic, chemical, or physical agents (including ionizing and nonionizing radiation) should be qualified to assess dangers associated with the programs and to select safeguards appropriate to the risks. An effective occupational health and safety program ensures that the risks associated with the experimental use of animals are reduced to acceptable levels. Potential hazards—such as animal bites, chemical cleaning agents, allergens, and zoonoses—that are inherent in or intrinsic to animal use should also be identified and evaluated. Health and safety specialists with knowledge in appropriate disciplines should be involved in the assessment of risks associated with hazardous activities and in the development of procedures to manage such risks. The extent and level of participation of personnel in the occupational health and safety program should be based on the hazards posed by the animals and materials used; on the exposure intensity, duration, and frequency; on the susceptibility of the personnel; and on the history of occupational illness and injury in the particular workplace (Clark 1993).

Personnel Training

Personnel at risk should be provided with clearly defined procedures for conducting their duties, should understand the hazards involved, and should be proficient in implementing the required safeguards.

Personnel should be trained regarding zoonoses, chemical safety, microbiologic and physical hazards (including those related to radiation and allergies), unusual conditions or agents that might be part of experimental procedures (including the use of genetically engineered animals and the use of human tissue in immunocompromised animals), handling of waste materials, personal hygiene, and other considerations (e.g., precautions to be taken during personnel pregnancy, illness, or decreased immunocompetence) as appropriate to the risk imposed by their workplace.

Personal Hygiene

It is essential that all personnel maintain a high standard of personal cleanliness. Clothing suitable for use in the animal facility and laboratories in which animals are used should be supplied and laundered by the institution. A commercial laundering service is acceptable in many situations; however, appropriate arrangements should be made to decontaminate clothing exposed to potential hazards. Disposable gloves, masks, head covers, coats, coveralls, and shoe covers might be desirable in some circumstances. Personnel should wash their hands and change clothing as often as necessary to maintain personal hygiene. Outer garments worn in the animal rooms should not be worn outside the animal facility. Personnel should not be permitted to eat, drink, use tobacco products, or apply cosmetics in animal rooms.

Facilities, Procedures, and Monitoring

Facilities required to support occupational health and safety concerns associated with animal care and use programs will vary. Because a high standard of personal cleanliness is essential, facilities and supplies for meeting this obligation should be provided. Washing and showering facilities appropriate to the program should be available. Facilities, equipment, and procedures should also be designed, selected, and developed to provide for ergonomically sound operations that reduce the potential of physical injury to personnel (such as might be caused by the lifting of heavy equipment or animals and the use of repetitive movements). Safety equipment should be properly maintained and routinely calibrated.

The selection of appropriate animal-housing systems requires professional knowledge and judgment and depends on the nature of the hazards in question, the types of animals used, and the design of the experiments. Experimental animals should be housed so that potentially contaminated food and bedding, feces, and urine can be handled in a controlled manner. Facilities, equipment, and procedures should be provided for appropriate bedding disposal.

Appropriate methods should be used for assessing exposure to potentially hazardous biologic, chemical, and physical agents where the possibility of exceeding permissible exposure limits (PELs) exists (CFR 1984b).

Animal Experimentation Involving Hazards

In selecting specific safeguards for animal experimentation with hazardous agents, careful attention should be given to procedures for animal care and housing, storage and disbursement of the agents, dose preparation and administration, body-fluid and tissue handling, waste and carcass disposal, and personal protection. Special safety equipment should be used in combination with appropriate management and safe practices. As a general rule, safety depends on trained personnel who rigorously follow safe practices.

Institutions should have written policies governing experimentation with hazardous biologic, chemical, and physical agents. An oversight process (such as use of a safety committee) should be developed to involve persons who are knowledgeable in the evaluation of hazards and safety issues. Because the use of animals in such studies requires special considerations, the procedures and facilities to be used should undergo review for specific safety concerns. Formal safety programs should be established to assess the hazards, determine the safeguards needed for their control, ensure that the staff has the necessary training and skills, and ensure that the facilities are adequate for the safe conduct of the research. Technical support should be provided to monitor and ensure compliance with institutional safety policies.

The Centers for Disease Control and Prevention (CDC) and National Institutes of Health (NIH) publication *Biosafety in Microbiological and Biomedical Laboratories* (1993) and the National Research Council (In press) recommend practices and procedures, safety equipment, and facility requirements for working with hazardous biologic agents and materials. Facilities that handle agents of unknown risk should consult with appropriate CDC personnel about hazard control and medical surveillance.

Special facilities and safety equipment are needed to protect the animal-care and investigative staff, other occupants of the facility, the public, animals, and the environment from exposure to hazardous biologic, chemical, and physical agents used in animal experimentation. Facilities used for animal experimentation with hazardous agents should be separated from other animal housing and support areas, research and clinical laboratories, and patient-care facilities and should be appropriately identified; and access to them should be limited to authorized personnel. Such facilities should be designed and constructed to facilitate cleaning and maintenance of mechanical systems. A properly managed and used double-corridor facility or barrier entry system is an effective means of reducing cross-contamination. Floor drains should always contain liquid or be sealed effectively by other means. Automatic trap priming can be provided to ensure that traps remain filled.

Hazardous agents should be contained within the study environment. Control of airflow (such as through the use of biologic-safety cabinets) that minimizes the escape of contaminants is a primary barrier used in the handling and administration of hazardous agents and the performance of necropsies on contaminated animals (CDC 1995; Kruse and others 1991). Special features of the facility—such as airlocks, negative air pressure, air filters, and redundant mechanical equipment with automatic switching—are secondary barriers aimed at preventing accidental release of hazards outside the facility and work environment.

Exposure to anesthetic waste gases should be limited. This is usually accomplished by using various scavenging techniques. If ether is used, personnel safety should be ensured by proper use of signs and by using equipment and practices to minimize risks associated with its explosiveness.

Personal Protection

Personal protective equipment should be provided, and other safety measures should be adopted when needed. Animal-care personnel should wear appropriate institution-issued protective clothing, shoes or shoe covers, and gloves. Clean protective clothing should be provided as often as necessary. If it is appropriate, personnel should shower when they leave the animal-care, procedure, or dose-preparation areas. Protective clothing and equipment should not be worn beyond the boundary of the hazardous-agent work area or the animal facility. Personnel with potential exposure to hazardous agents should be provided with personal protective equipment appropriate to the agents (CFR 1984c). For example, personnel exposed to nonhuman primates should be provided with such protective items as gloves, arm protectors, masks, and face shields. Hearing protection should be provided in high-noise areas. Personnel working in areas where they might be exposed to contaminated airborne particulate material or vapors should be provided with suitable respiratory protection (CFR 1984c).

Medical Evaluation and Preventive Medicine for Personnel

Development and implementation of a program of medical evaluation and preventive medicine should involve input from trained health professionals, such as occupational-health physicians and nurses. Confidentiality and other medical and legal factors must be considered in the context of appropriate federal, state, and local regulations.

A health-history evaluation before work assignment is advisable to assess potential risks for individual employees. Periodic medical evaluations are advisable for people in some risk categories. An appropriate immunization schedule should be adopted. It is important to immunize animal-care personnel against tetanus. In addition, pre-exposure immunization should be offered to people at risk of infection or exposure to such agents as rabies or hepatitis B virus. Vaccination is recommended if research is to be conducted on infectious diseases for which effective vaccines are available. Specific recommendations can be found in the CDC and NIH publication *Biosafety in Microbiological and Biomedical Laboratories* (1993). Pre-employment or pre-exposure serum collection is advisable only in specific circumstances as determined by an occupational health and safety professional (NRC In press). In such cases, identification, traceability, retention, and storage conditions of samples should be considered, and the purpose for which the serum samples will be used must be consistent with applicable state laws and consistent with the Federal Policy for the Protection of Human Subjects (Federal Register 56(117): 28002-28032, June 18, 1991).

Zoonosis surveillance should be a part of an occupational-health program (CDC and NIH 1993; Fox and others 1984; NRC In press). Personnel should be instructed to notify their supervisors of potential or known exposures and of suspected health hazards and illnesses. Clear procedures should be established for reporting all accidents, bites, scratches, and allergic reactions (NRC In press).

Nonhuman-primate diseases that are transmissible to humans can be serious hazards. Animal technicians, clinicians, investigators, predoctoral and postdoctoral trainees, research technicians, consultants, maintenance workers, security personnel, and others who have contact with nonhuman primates or have duties in nonhuman-primate housing areas should be routinely screened for tuberculosis. Because of the potential for *Cercopithecine herpesvirus 1* (formerly *Herpesvirus simiae*) exposure, personnel who work with macaques should have access to and be instructed in the use of bite and scratch emergency-care stations (Holmes and others 1995). A procedure should be established for ensuring medical care for bites and scratches.

REFERENCES

AVMA (American Veterinary Medical Association). 1995. Accredited programs in veterinary technology. Pp. 236-240 in 1995 AVMA Membership Directory and Resource Manual, 44th ed. Schaumburg, Ill.: AVMA.

Bryant, J. M. 1980. Vest and tethering system to accommodate catheters and a temperature monitor for nonhuman primates. Lab. Anim. Sci. 30(4, Part I):706-708.

Byrd, L. D. 1979. A tethering system for direct measurement of cardiovascular function in the caged baboon. Am. J. Physiol. 236:H775-H779.

CCAC (Canadian Council on Animal Care) 1993. Guide to the Care and Use of Experimental Animals, Vol. 1, 2nd ed. E. D. Olfert, B. M. Cross, and A. A. McWilliam, eds. Ontario, Canada: Canadian Council on Animal Care. 211 pp.

CDC (Centers for Disease Control and Prevention and NIH (National Institutes of Health). 1993. Biosafety in Microbiological and Biomedical Laboratories. 3rd ed. HHS Publication No. (CDC) 93-8395, Washington, D.C.: U.S. Government Printing Office.

CDC (Centers for Disease Control and Prevention) and NIH (National Institutes of Health). 1995. Primary Containment for Biohazards: Selection, Installation and Use of Biological Safety Cabinets. Washington, D.C.: U.S. Government Printing Office.

CFR (Code of Federal Regulations). 1984a. Title 10; Part 20, Standards for Protection Against Radiation. Washington, D.C.: Office of the Federal Register.

CFR (Code of Federal Regulations). 1984b. Title 29; Part 1910, Occupational Safety and Health Standards; Subpart G, Occupation Health and Environmental Control, and Subpart Z, Toxic and Hazardous Substances. Washington, D.C.: Office of the Federal Register.

CFR (Code of Federal Regulations). 1984c. Title 29; Part 1910, Occupational Safety and Health Standards; Subpart I, Personal Protective Equipment. Washington, D.C.: Office of the Federal Register.

CFR (Code of Federal Regulations). 1985. Title 9 (Animals and Animal Products), Subchapter A (Animal Welfare). Washington, D.C.: Office of the Federal Register.

Clark, J. M. 1993. Planning for safety: biological and chemical hazards. Lab Anim. 22:33-38.

Fox, J. G., C. E. Newcomer, and H. Rozmiarek. 1984. Selected zoonoses and other health hazards. Pp. 614-648 in Laboratory Animal Medicine, J. G. Fox, B. J. Cohen, and F. M. Loew, eds. New York: Academic Press.

Grandin, T. 1991. Livestock behavior and the design of livestock handling facilities. Pp. 96-125 in Handbook of Facilities Planning. Vol. 2. Laboratory Animal Facilities. New York: Van Nostrand. 422 pp.

Holmes, G. P., L. E. Chapman, J. A. Stewart, S. E. Straus, J. K. Hilliard, D. S. Davenport, and the B Virus Working Group. 1995. Guidelines for the prevention and treatment of B-virus infections in exposed persons. Clin. Infect. Dis. 20:421-439.

IRAC (Interagency Research Animal Committee). 1985. U.S. Government Principles for Utilization and Care of Vertebrate Animals Used in Testing, Research, and Training. Federal Register, May 20, 1985. Washington, D.C.: Office of Science and Technology Policy.

Kreger, M. D., 1995. Training Materials for Animal Facility Personnel: AWIC Quick Bibliography Series, 95-08. Beltsville, Md.: National Agricultural Library.

Kruse, R. H., W. H. Puckett, and J. H. Richardson. 1991. Biological safety cabinetry. Clin. Micro. Reviews 4:207-241.

McNamee, G. A., Jr., R. W. Wannemacher, Jr., R. E. Dinterman, H. Rozmiarek, and R. D. Montrey. 1984. A surgical procedure and tethering system for chronic blood sampling, infusion, and temperature monitoring in caged nonhuman primates. Lab. Anim. Sci. 34(3):303-307.

Morton, W. R., G. H. Knitter, P. M. Smith, T. G. Susor, and K. Schmitt. 1987. Alternatives to chronic restraint of nonhuman primates. J. Am. Vet. Med. Assoc. 191(10):1282-1286.

NIH (National Institutes of Health). 1990. Guidelines for Diet Control in Behavioral Study. Bethesda, Md.: Animal Research Advisory Committee, NIH.

NRC (National Research Council). 1990. Important laboratory animal resources: selection criteria and funding mechanisms for their preservation. A report of the Institute of Laboratory Animal Resources Committee on Preservation of Laboratory Animal Resources. ILAR News 32(4):A1-A32.

NRC (National Research Council). 1991. Education and Training in the Care and Use of Laboratory Animals: A Guide for Developing Institutional Programs. A report of the Institute of Laboratory Animal Resources Committee on Educational Programs in Laboratory Animal Science. Washington, D.C.: National Academy Press. 152 pp.

NRC (National Research Council). In press. Occupational Health and Safety in the Care and Use of Research Animals. A report of the Institute of Laboratory Animal Resources Committee on Occupational Safety and Health in Research Animal Facilities. Washington, D.C.: National Academy Press.

NYAS (New York Academy of Sciences). 1988. Interdisciplinary Principles and Guidelines for the Use of Animals in Research, Testing and Education. New York: New York Academy of Sciences.

PHS (Public Health Service). 1996. Public Health Service Policy on Humane Care and Use of Laboratory Animals. Washington, D.C.: U.S. Department of Health and Human Services, 28 pp. [PL 99-158, Health Research Extension Act, 1985]

Reinhardt, V. 1991. Training adult male rhesus monkeys to actively cooperate during in-homecage venipuncture. Anim. Technol. 42(1):11-17.

Reinhardt, V. 1995. Restraint methods of laboratory non-human primates: a critical review. Anim. Welf. 4:221-238.

Van Sluyters, R. C., and M. D. Oberdorfer, eds. 1991. Preparation and Maintenance of Higher Mammals During Neuroscience Experiments. Report of National Institute of Health Workshop. NIH No. 91-3207. Bethesda, Md.: National Institutes of Health.

Wakeley, H., J. Dudek, and J. Kruckeberg. 1974. A method for preparing and maintaining rhesus monkeys with chronic venous catheters. Behav. Res. Methods Instrum. 6:329-331.

2

Animal Environment, Housing, and Management

Proper housing and management of animal facilities are essential to animal well-being, to the quality of research data and teaching or testing programs in which animals are used, and to the health and safety of personnel. A good management program provides the environment, housing, and care that permit animals to grow, mature, reproduce, and maintain good health; provides for their well-being; and minimizes variations that can affect research results. Specific operating practices depend on many factors that are peculiar to individual institutions and situations. Well-trained and motivated personnel can often ensure high-quality animal care, even in institutions with less than optimal physical plants or equipment.

Many factors should be considered in planning for adequate and appropriate physical and social environment, housing, space, and management. These include

- The species, strain, and breed of the animal and individual characteristics, such as sex, age, size, behavior, experiences, and health.
- The ability of the animals to form social groups with conspecifics through sight, smell, and possibly contact, whether the animals are maintained singly or in groups.
- The design and construction of housing.
- The availability or suitability of enrichments.
- The project goals and experimental design (e.g., production, breeding, research, testing, and teaching).

- The intensity of animal manipulation and invasiveness of the procedures conducted.
- The presence of hazardous or disease-causing materials.
- The duration of the holding period.

Animals should be housed with a goal of maximizing species-specific behaviors and minimizing stress-induced behaviors. For social species, this normally requires housing in compatible pairs or groups. A strategy for achieving desired housing should be developed by animal-care personnel with review and approval by the IACUC. Decisions by the IACUC, in consultation with the investigator and veterinarian, should be aimed at achieving high standards for professional and husbandry practices considered appropriate for the health and well-being of the species and consistent with the research objectives. After the decision-making process, objective assessments should be made to substantiate the adequacy of animal environment, husbandry, and management.

The environment in which animals are maintained should be appropriate to the species, its life history, and its intended use. For some species, it might be appropriate to approximate the natural environment for breeding and maintenance. Expert advice might be sought for special requirements associated with the experiment or animal subject (for example, hazardous-agent use, behavioral studies, and immunocompromised animals, farm animals, and nontraditional laboratory species).

The following sections discuss some considerations of the physical environment related to common research animals.

PHYSICAL ENVIRONMENT

Microenvironment and Macroenvironment

The *microenvironment* of an animal is the physical environment immediately surrounding it—the primary enclosure with its own temperature, humidity, and gaseous and particulate composition of the air. The physical environment of the secondary enclosure—such as a room, a barn, or an outdoor habitat—constitutes the *macroenvironment*. Although the microenvironment and the macroenvironment are linked by ventilation between the primary and secondary enclosures, the environment in the primary enclosure can be quite different from the environment in the secondary enclosure and is affected by the design of both enclosures.

Measurement of the characteristics of the microenvironment can be difficult in small primary enclosures. Available data indicate that temperature, humidity, and concentrations of gases and particulate matter are often higher in an animal's microenvironment than in the macroenvironment (Besch 1980; Flynn 1959; Gamble and Clough 1976; Murakami 1971; Serrano 1971). Microenvironmental conditions can induce changes in metabolic and physiologic processes or alterations in disease susceptibility (Broderson and others 1976; Schoeb and others 1982; Vesell and others 1976).

Housing

Primary Enclosures

The primary enclosure (usually a cage, pen, or stall) provides the limits of an animal's immediate environment. Acceptable primary enclosures

- Allow for the normal physiologic and behavioral needs of the animals, including urination and defecation, maintenance of body temperature, normal movement and postural adjustments, and, where indicated, reproduction.
- Allow conspecific social interaction and development of hierarchies within or between enclosures.
- Make it possible for the animals to remain clean and dry (as consistent with the requirements of the species).
- Allow adequate ventilation.
- Allow the animals access to food and water and permit easy filling, refilling, changing, servicing, and cleaning of food and water utensils.
- Provide a secure environment that does not allow escape of or accidental entrapment of animals or their appendages between opposing surfaces or by structural openings.
- Are free of sharp edges or projections that could cause injury to the animals.
- Allow observation of the animals with minimal disturbance of them.

Primary enclosures should be constructed with materials that balance the needs of the animal with the ability to provide for sanitation. They should have smooth, impervious surfaces with minimal ledges, angles, corners, and overlapping surfaces so that accumulation of dirt, debris, and moisture is reduced and satisfactory cleaning and disinfecting are possible. They should be constructed of durable materials that resist corrosion and withstand rough handling without chipping, cracking, or rusting. Less-durable materials, such as wood, can provide a more appropriate environment in some situations (such as runs, pens, and outdoor corrals) and can be used to construct perches, climbing structures, resting areas, and perimeter fences for primary enclosures. Wooden items might need to be replaced periodically because of damage or difficulties with sanitation.

All primary enclosures should be kept in good repair to prevent escape of or injury to animals, promote physical comfort, and facilitate sanitation and servicing. Rusting or oxidized equipment that threatens the health or safety of the animals should be repaired or replaced.

Some housing systems have special caging and ventilation equipment, including filter-top cages, ventilated cages, isolators, and cubicles. Generally, the purpose of these systems is to minimize the spread of airborne disease agents between cages or groups of cages. They often require different husbandry practices, such as alterations in the frequency of bedding change, the use of aseptic handling techniques, and specialized cleaning, disinfecting, or sterilization regimens to prevent microbial transmission by other than the airborne route.

Rodents are often housed on wire flooring, which enhances sanitation of the cage by enabling urine and feces to pass through to a collection tray. However, some evidence suggests that solid-bottom caging, with bedding, is preferred by rodents (Fullerton and Gilliatt 1967; Grover-Johnson and Spencer 1981; Ortman and others 1983). Solid-bottom caging, with bedding, is therefore recommended for rodents. Vinyl-coated flooring is often used for other species, such as dogs and nonhuman primates. IACUC review of this aspect of the animal care program should ensure that caging enhances animal well-being consistent with good sanitation and the requirements of the research project.

Sheltered or Outdoor Housing

Sheltered or outdoor housing—such as barns, corrals, pastures, and islands—is a common primary housing method for some species and is acceptable for many situations. In most cases, outdoor housing entails maintaining animals in groups.

When animals are maintained in outdoor runs, pens, or other large enclosures, there must be protection from extremes in temperature or other harsh weather conditions and adequate protective and escape mechanisms for submissive animals. These goals can be achieved by such features as windbreaks, shelters, shaded areas, areas with forced ventilation, heat-radiating structures, or means of retreat to conditioned spaces, such as an indoor portion of a run. Shelters should be accessible to all animals, have sufficient ventilation, and be designed to prevent buildup of waste materials and excessive moisture. Houses, dens, boxes, shelves, perches, and other furnishings should be constructed in a manner and made of materials that allow cleaning or replacement in accord with generally accepted husbandry practices when the furnishings are excessively soiled or worn.

Floors or ground-level surfaces of outdoor housing facilities can be covered with dirt, absorbent bedding, sand, gravel, grass, or similar material that can be removed or replaced when that is needed to ensure appropriate sanitation. Excessive buildup of animal waste and stagnant water should be avoided by, for example, using contoured or drained surfaces. Other surfaces should be able to withstand the elements and be easily maintained.

Successful management of outdoor housing relies on consideration of

• An adequate acclimation period in advance of seasonal changes when animals are first introduced to outdoor housing.
• Training of animals to cooperate with veterinary and investigative personnel and to enter chutes or cages for restraint or transport.
• Species-appropriate social environment.
• Grouping of compatible animals.
• Adequate security via a perimeter fence or other means.

Naturalistic Environments

Areas like pastures and islands afford opportunities to provide a suitable environment for maintaining or producing animals and for some types of research. Their use results in the loss of some control over nutrition, health care and surveillance, and pedigree management. These limitations should be balanced against the benefits of having the animals live in more natural conditions. Animals should be added to, removed from, and returned to social groups in this setting with appropriate consideration of the effects on the individual animals and on the group. Adequate supplies of food, fresh water, and natural or constructed shelter should be ensured.

Space Recommendations

An animal's space needs are complex, and consideration of only the animal's body weight or surface area is insufficient. Therefore, the space recommendations presented here are based on professional judgment and experience and should be considered as recommendations of appropriate cage sizes for animals under conditions commonly found in laboratory animal housing facilities. Vertical height, structuring of the space, and enrichments can clearly affect animals' use of space. Some species benefit more from wall space (e.g., "thigmotactic" rodents), shelters (e.g., some New World primates), or cage complexities (e.g., cats and chimpanzees) than from simple increases in floor space (Anzaldo and others 1994; Stricklin 1995). Thus, basing cage-size recommendations on floor space alone is inadequate. In this regard, the *Guide* might differ from the AWRs (see footnote 1, p. 2).

Space allocations should be reviewed and modified as necessary to address individual housing situations and animal needs (for example, for prenatal and postnatal care, obese animals, and group or individual housing). Such animal-performance indexes as health, reproduction, growth, behavior, activity, and use of space can be used to assess the adequacy of housing. At a minimum, an animal must have enough space to turn around and to express normal postural adjustments, must have ready access to food and water, and must have enough clean-bedded or unobstructed area to move and rest in. For cats, a raised resting surface should be included in the cage. Raised resting surfaces or perches are also often desirable for dogs and nonhuman primates. Low resting surfaces that do not allow the space under them to be comfortably occupied by the animal should be counted as part of the floor space. Floor space taken up by food bowls, water containers, litter boxes, or other devices not intended for movement or resting should not be considered part of the floor space.

The need for and type of adjustments in the amounts of primary enclosure space recommended in the tables that follow should be approved at the institutional level by the IACUC and should be based on the performance outcomes described in the preceding paragraph with due consideration of the AWRs and PHS Policy (see footnote 1, p. 2). Professional judgment, surveys of the literature and current practices, and consideration of the animals' physical, behavioral, and social needs and of the nature of the protocol and its requirements might be necessary (see Crockett and others 1993, 1995). Assessment of animals' space needs should be a continuing process. With the passage of time or long-term protocols, adjustments in floor space and height should be considered and modified as necessary.

It is not within the scope or size constraints of the *Guide* to discuss the housing requirements of all species used in research. For species not mentioned, space and height allocations for an animal of equivalent size and with a similar activity profile and similar behavior can be used as a starting point from which adjustments that take species-specific and individual needs into account can be made.

Whenever it is appropriate, social animals should be housed in pairs or groups, rather than individually, provided that such housing is not contraindicated by the protocol in question and does not pose an undue risk to the animals (Brain and Bention 1979). Depending on a variety of biologic and behavioral factors, group-housed animals might need less or more total space per animal than individually housed animals. Recommendations provided below are based on the assumption that pair or group housing is generally preferable to single housing, even when members of the pair or group have slightly less space *per animal* than when singly caged. For example, each animal can share the space allotted to the animals with which it is housed. Furthermore, some rodents or swine housed in compatible groups seek each other out and share cage space by huddling together along walls, lying on each other during periods of rest, or gathering in areas of retreat (White 1990; White and others 1989). Cattle, sheep, and goats exhibit herding behavior and seek group associations and close physical contact. Conversely, some animals, such as various species of nonhuman primates, might need additional individual space when group-housed to reduce the level of aggression.

The height of enclosures can be important in the normal behavior and postural adjustments of some species. Cage heights should take into account typical postures of an animal and provide adequate clearance for normal cage components, such as feeders and water devices, including sipper tubes. Some species of nonhuman primates use the vertical dimensions of the cage to a greater extent than the floor. For them, the ability to perch and to have adequate vertical space to keep the whole body above the cage floor can improve their well-being.

Space allocations for animals should be based on the following tables, but might need to be increased, or decreased with approval of the IACUC, on the basis of criteria previously listed.

Table 2.1 lists recommended space allocations for commonly used laboratory rodents housed in groups. If they are housed individually or exceed the weights in the table, animals might require more space.

Table 2.2 lists recommended space allocations for other common laboratory animals. These allocations are based, in general, on the needs of individually housed animals. Space allocations should be re-evaluated to provide for enrichment of the primary enclosure or to accommodate animals that exceed the weights in the table. For group housing, determination of the total space needed is not necessarily based on the sum of the amounts recommended for individually housed animals. Space for group-housed animals should be based on individual species needs, behavior, compatibility of the animals, numbers of animals, and goals of the housing situation.

TABLE 2.1 Recommended Space for Commonly Used Group-Housed Laboratory Rodents

Animals	Weight, g	Floor Area/Animal, in^2 [a]	Height,[b] in[c]
Mice	<10	6	5
	Up to 15	8	5
	Up to 25	12	5
	>25[d]	>15	5
Rats	<100	17	7
	Up to 200	23	7
	Up to 300	29	7
	Up to 400	40	7
	Up to 500	60	7
	>500[d]	>70	7
Hamsters	<60	10	6
	Up to 80	13	6
	Up to 100	16	6
	>100[d]	>19	6
Guinea pigs	≤350	60	7
	>350[d]	≥101	7

[a]To convert square inches to square centimeters, multiply by 6.45.
[b]From cage floor to cage top.
[c]To convert inches to centimeters, multiply by 2.54.
[d]Larger animals might require more space to meet the performance standards (see text).

TABLE 2.2   Recommended Space for Rabbits, Cats, Dogs, Nonhuman Primates, and Birds

Animals	Weight, kg[a]	Floor Area/Animal, ft^2 [b]	Height [c] in[d]
Rabbits	<2	1.5	14
	Up to 4	3.0	14
	Up to 5.4	4.0	14
	>5.4[e]	>5.0	14
Cats	≤4	3.0	24
	>4[e]	≥4.0	24
Dogs[f]	<15	8.0	—
	Up to 30	12.0	—
	>30[e]	≥24.0	—
Monkeys[g, h] (including baboons)			
Group 1	Up to 1	1.6	20
Group 2	Up to 3	3.0	30
Group 3	Up to 10	4.3	30
Group 4	Up to 15	6.0	32
Group 5	Up to 25	8.0	36
Group 6	Up to 30	10.0	46
Group 7	>30[e]	15.0	46
Apes (Pongidae)[h]			
Group 1	Up to 20	10.0	55
Group 2	Up to 35	15.0	60
Group 3	>35[i]	25.0	84
Pigeons[j]	—	0.8	—
Quail[j]	—	0.25	—
Chickens[j]	<0.25	0.25	—
	Up to 0.5	0.50	—
	Up to 1.5	1.00	—
	Up to 3.0	2.00	—
	>3.0[e]	≥3.00	—

Table 2.3 lists recommended space allocations for farm animals commonly used in a laboratory setting. When animals, housed individually or in groups, exceed the weights in the table, more space might be required. If they are group-housed, adequate access to water and feeder space should be provided (Larson and Hegg 1976; Midwest Plan Service 1987).

Temperature and Humidity

Regulation of body temperature within normal variation is necessary for the well-being of homeotherms. Generally, exposure of unadapted animals to temperatures above 85°F (29.4°C) or below 40°F (4.4°C), without access to shelter or other protective mechanisms, might produce clinical effects (Gordon 1990),

TABLE 2.2 Continued

[a]To convert kilograms to pounds, multiply by 2.2.

[b]To convert square feet to square meters, multiply by 0.09.

[c]From cage floor to cage top.

[d]To convert inches to centimeters, multiply by 2.54.

[e]Larger animals might require more space to meet performance standards (see text).

[f]These recommendations might require modification according to body conformation of individual animals and breeds. Some dogs, especially those toward upper limit of each weight range, might require additional space to ensure compliance with the regulations of the Animal Welfare Act. These regulations (CFR 1985) mandate that the height of each cage be sufficient to allow occupant to stand in "comfortable position" and that the minimal square feet of floor space be equal to "mathematical square of the sum of the length of the dog in inches (measured from the tip of its nose to the base of its tail) plus 6 inches; then divide the product by 144."

[g]Callitrichidae, Cebidae, Cercopithecidae, and *Papio*. Baboons might require more height than other monkeys.

[h]For some species (e.g., *Brachyteles*, *Hylobates*, *Symphalangus*, *Pongo*, and *Pan*), cage height should be such that an animal can, when fully extended, swing from the cage ceiling without having its feet touch the floor. Cage-ceiling design should enhance brachiating movement.

[i]Apes weighing over 50 kg are more effectively housed in permanent housing of masonry, concrete, and wire-panel structure than in conventional caging.

[j]Cage height should be sufficient for the animals to stand erect with their feet on the floor.

which could be life-threatening. Animals can adapt to extremes by behavioral, physiologic, and morphologic mechanisms, but such adaptation takes time and might alter protocol outcomes or otherwise affect performance (Garrard and others 1974; Gordon 1993; Pennycuik 1967).

Environmental temperature and relative humidity can depend on husbandry and housing design and can differ considerably between primary and secondary enclosures. Factors that contribute to variation in temperature and humidity include housing material and construction, use of filter tops, number of animals per cage, forced ventilation of the enclosures, frequency of bedding changes, and bedding type.

Some conditions might require increased environmental temperatures, such as postoperative recovery, maintenance of chicks for the first few days after hatching, housing of some hairless rodents, and housing of neonates that have been separated from their mothers. The magnitude of the temperature increase depends on the circumstances of housing; sometimes, raising the temperature in the primary enclosure alone (rather than raising the temperature of the secondary enclosure) is sufficient.

In the absence of well-controlled studies, professional judgment and experience have resulted in recommendations for dry-bulb temperatures (Table 2.4) for several common species. In the case of animals in confined spaces, the range of

TABLE 2.3 Recommended Space for Commonly Used Farm Animals

Animals/Enclosure	Weight, kg[a]	Floor Area/Animal, ft^2 [b]
Sheep and Goats		
1	<25	10.0
	Up to 50	15.0
	>50[c]	20.0
2-5	<25	8.5
	Up to 50	12.5
	>50[c]	17.0
>5	<25	7.5
	Up to 50	11.3
	>50[c]	15.0
Swine		
1	<15	8.0
	Up to 25	12.0
	Up to 50	15.0
	Up to 100	24.0
	Up to 200	48.0
	>200[c]	>60.0
2-5	<25	6.0
	Up to 50	10.0
	Up to 100	20.0
	Up to 200	40.0
	>200[c]	≥52.0
>5	<25	6.0
	Up to 50	9.0
	Up to 100	18.0
	Up to 200	36.0
	>200[c]	≥48.0

daily temperature fluctuations should be kept to a minimum to avoid repeated large demands on the animals' metabolic and behavioral processes to compensate for changes in the thermal environment. Relative humidity should also be controlled, but not nearly as narrowly as temperature; the acceptable range of relative humidity is 30 to 70%. The temperature ranges in Table 2.4 might not apply to captive wild animals, wild animals maintained in their natural environment, or animals in outdoor enclosures that are given the opportunity to adapt by being exposed to seasonal changes in ambient conditions.

Ventilation

The purposes of ventilation are to supply adequate oxygen; remove thermal loads caused by animal respiration, lights, and equipment; dilute gaseous and particulate contaminants; adjust the moisture content of room air; and, where

TABLE 2.3 Continued

Animals/Enclosure	Weight, kg[a]	Floor Area/Animal, ft^2 [b]
Cattle		
1	<75	24.0
	Up to 200	48.0
	Up to 350	72.0
	Up to 500	96.0
	Up to 650	124.0
	>650[c]	>144.0
2-5	<75	20.0
	Up to 200	40.0
	Up to 350	60.0
	Up to 500	80.0
	Up to 650	105.0
	>650[c]	>120.0
>5	<75	18.0
	Up to 200	36.0
	Up to 350	54.0
	Up to 500	72.0
	Up to 650	93.0
	>650[c]	>108.0
Horses	—	144.0
Ponies		
1-4	—	72.0
>4/Pen	≤200	60.0
	>200[c]	>72.0

[a]To convert kilograms to pounds, multiply by 2.2.
[b]To convert square feet to square meters, multiply by 0.09.
[c]Larger animals might require more space to meet performance standards (see text).

appropriate, create static-pressure differentials between adjoining spaces. Establishing a room ventilation rate, however, does not ensure the adequacy of the ventilation of an animal's primary enclosure and hence does not guarantee the quality of the microenvironment.

The degree to which air movement (drafts) causes discomfort or biologic consequences has not been established for most species. The volume and physical characteristics of the air supplied to a room and its diffusion pattern influence the ventilation of an animal's primary enclosure and so are important determinants of its microenvironment. The relationship of the type and location of supply-air diffusers and exhaust vents to the number, arrangement, location, and type of primary enclosures in a room or other secondary enclosure affects how well the primary enclosures are ventilated and should therefore be considered. The use of computer modeling for assessing those factors in relation to heat loading and air diffusion patterns can be helpful in optimizing ventilation of primary and

TABLE 2.4   Recommended Dry-Bulb Temperatures for Common Laboratory
Animals

Animal	Dry-Bulb Temperature	
	°C	°F
Mouse, rat, hamster, gerbil, guinea pig	18-26	64-79
Rabbit	16-22	61-72
Cat, dog, nonhuman primate	18-29	64-84
Farm animals and poultry	16-27	61-81

secondary enclosures (for example, Hughes and Reynolds 1995; Reynolds and Hughes 1994).

The guideline of 10-15 fresh-air changes per hour has been used for secondary enclosures for many years and is considered an acceptable general standard. Although it is effective in many animal-housing settings, the guideline does not take into account the range of possible heat loads; the species, size, and number of animals involved; the type of bedding or frequency of cage-changing; the room dimensions; or the efficiency of air distribution from the secondary to the primary enclosure. In some situations, the use of such a broad guideline might pose a problem by overventilating a secondary enclosure that contains few animals and thereby wasting energy or by underventilating a secondary enclosure that contains many animals and thereby allowing heat and odor accumulation.

To determine more accurately the ventilation required, the minimal ventilation rate (commonly in cubic feet per minute) required to accommodate heat loads generated by animals can be calculated with the assistance of mechanical engineers. The heat generated by animals can be calculated with the average-total-heat-gain formula as published by the American Society of Heating, Refrigeration, and Air-Conditioning Engineers (ASHRAE, 1993). The formula is species-independent, so it is applicable to any heat-generating animal. Minimal required ventilation is determined by calculating the amount of cooling required (total cooling load) to control the heat load expected to be generated by the largest number of animals to be housed in the enclosure in question plus any heat expected to be produced by nonanimal sources and heat transfer through room surfaces. The total-cooling-load calculation method can also be used for an animal space that has a fixed ventilation rate to determine the maximal number of animals (based on total animal mass) that can be housed in the space.

Even though that calculation can be used to determine minimal ventilation needed to prevent heat buildup, other factors—such as odor control, allergen control, particle generation, and control of metabolically generated gases—might necessitate ventilation beyond the calculated minimum. When the calculated minimal required ventilation is substantially less than 10 air changes per hour, lower ventilation rates might be appropriate in the secondary enclosure, provided that they do not result in harmful or unacceptable concentrations of toxic gases, odors, or particles in the primary enclosure. Similarly, when the calculated minimal required ventilation exceeds 15 air changes per hour, provisions should be made for additional ventilation required to address the other factors. In some cases, fixed ventilation in the secondary enclosure might necessitate adjustment of sanitation schedules or limitation of animal numbers to maintain appropriate environmental conditions.

Caging with forced ventilation that uses filtered room air and other types of special primary enclosures with independent air supplies (i.e., air not drawn from the room) can effectively address the ventilation requirements of animals without the need to ventilate secondary enclosures to the extent that would be needed if there were no independent primary-enclosure ventilation. Nevertheless, a secondary enclosure should be ventilated sufficiently to provide for the heat loads released from its primary enclosures. If the specialized enclosures contain adequate particulate and gaseous filtration to address contamination risks, recycled air may be used in the secondary enclosures.

Filtered isolation caging without forced ventilation, such as that used in some types of rodent housing, restricts ventilation. To compensate, it might be necessary to adjust husbandry practices—including sanitation, placement of cages in the secondary enclosure, and cage densities—to improve the microenvironment and heat dissipation.

The use of recycled air to ventilate animal rooms saves considerable amounts of energy but might entail some risk. Many animal pathogens can be airborne or travel on fomites, such as dust, so exhaust air to be recycled into heating, ventilation, and air conditioning (HVAC) systems that serve multiple rooms presents a risk of cross contamination. The exhaust air to be recycled should be HEPA-filtered (high-efficiency particulate air-filtered) to remove airborne particles before it is recycled; the extent and efficiency of filtration should be proportional to the estimated risk. HEPA filters are available in various efficiencies that can be used to match the magnitude of risk (ASHRAE 1992, 1993). Air that does not originate from animal-use areas but has been used to ventilate other spaces (e.g., some human-occupancy areas and food, bedding, and supply storage areas) may be recycled for animal-space ventilation and might require less-intensive filtration or conditioning than air recycled from animal-use space. The risks in some situations, however, might be too great to consider recycling (e.g., in the case of nonhuman-primate and biohazard areas).

Toxic or odor-causing gases, such as ammonia, can be kept within acceptable limits if they are removed by the ventilation system and replaced with air that contains either a lower concentration or none of these gases. Treatment of recycled air for these substances by chemical absorption or scrubbing might be effective; however, the use of nonrecycled air is preferred for ventilation of animal use and holding areas. The use of HEPA-filtered recycled air without gaseous filtration (such as with activated-charcoal filters) can be used but only in limited applications, provided that

- Room air is mixed with at least 50% fresh air (that is, the supply air does not exceed 50% recycled air).
- Husbandry practices, such as bedding-change and cage-washing frequency, and the preparation of recycled air used are sufficient to minimize toxic gases and odors.
- Recycled air is returned only to the room or area from which it was generated, except if it comes from other than animal-housing areas.
- Recycled air is appropriately conditioned and mixed with sufficient fresh air to address the thermal and humidity requirements of animals in that space.

Frequent bedding changes and cage-cleaning coupled with husbandry practices, such as low animal density within the room and lower environmental temperature and humidity, can also reduce the concentration of toxic or odor-causing gases in animal-room air. Treatment of recycled air for either particulate or gaseous contaminants is expensive and can be rendered ineffective by improper or insufficient maintenance of filtration systems. These systems should be properly maintained and monitored appropriately to maximize their effectiveness.

The successful operation of any HVAC system requires regular maintenance and evaluation, including measurement of its function at the level of the secondary enclosure. Such measurements should include supply- and exhaust-air volumes, as well as static-pressure differentials, where applicable.

Illumination

Light can affect the physiology, morphology, and behavior of various animals (Brainard and others 1986; Erkert and Grober 1986; Newbold and others 1991; Tucker and others 1984). Potential photostressors include inappropriate photoperiod, photointensity, and spectral quality of the light (Stoskopf 1983). Numerous factors can affect animals' needs for light and should be considered when an appropriate illumination level is being established for an animal holding room. These include light intensity, duration of exposure, wavelength of light, light history of the animal, pigmentation of the animal, time of light exposure during the circadian cycle, body temperature, hormonal status, age, species, sex, and stock or strain of animal (Brainard 1989; Duncan and O'Steen 1985; O'Steen 1980; Saltarelli and Coppola 1979; Semple-Rowland and Dawson 1987; Wax 1977).

In general, lighting should be diffused throughout an animal holding area and provide sufficient illumination for the well-being of the animals and to allow good housekeeping practices, adequate inspection of animals—including the bottom-most cages in racks—and safe working conditions for personnel. Light in animal holding rooms should provide for adequate vision and for neuroendocrine regulation of diurnal and circadian cycles (Brainard 1989).

Photoperiod is a critical regulator of reproductive behavior in many species of animals (Brainard and others 1986; Cherry 1987) and can also alter body-weight gain and feed intake (Tucker and others 1984). Inadvertent light exposure during the dark cycle should be minimized or avoided. Because some species will not eat in low light or darkness, such illumination schedules should be limited to a duration that will not compromise the well-being of the animals. A time-controlled lighting system should be used to ensure a regular diurnal cycle, and timer performance should be checked periodically to ensure proper cycling.

The most commonly used laboratory animals are nocturnal. Because the albino rat is more susceptible to phototoxic retinopathy than other species, it has been used as a basis for establishing room illumination levels (Lanum 1979). Data for room light intensities for other animals, based on scientific studies, are not available. Light levels of about 325 lux (30 ft-candles) about 1.0 m (3.3 ft) above the floor appear to be sufficient for animal care and do not cause clinical signs of phototoxic retinopathy in albino rats (Bellhorn 1980), and levels up to 400 lux (37 ft-candles) as measured in an empty room 1 m from the floor have been found to be satisfactory for rodents if management practices are used to prevent retinal damage in albinos (Clough 1982). However, the light experience of an individual animal can affect its sensitivity to phototoxicity; light of 130-270 lux above the light intensity under which it was raised has been reported to be near the threshold of retinal damage in some individual albino rats according to histologic, morphometric, and electrophysiologic evidence (Semple-Rowland and Dawson 1987). Some guidelines recommend a light intensity as low as 40 lux at the position of the animal in midcage (NASA 1988). Young albino and pigmented mice prefer much-lower illumination than adults (Wax 1977), although potential retinal damage associated with housing these rodents at higher light levels is mostly reversible. Thus, for animals that have been shown to be susceptible to phototoxic retinopathy, light at the cage level should be between 130 and 325 lux.

Management practices, such as rotating cage position relative to the light source (Greenman and others 1982) or providing animals with ways to modify their own light exposure by behavioral means (e.g., via tunneling or hiding in a structure), can be used to reduce inappropriate light stimulation of animals. Provision of variable-intensity light controls might be considered as a means of ensuring that light intensities are consistent with the needs of animals and personnel working in animal rooms and with energy conservation. Such controls should have some form of vernier scale and a lockable setting and should not be used merely to turn room lighting on and off. The Illuminating Engineering Society of North America (IESNA) handbook (Kaufman 1984, 1987) can assist in decisions concerning lighting uniformity, color-rendering index, shielding, glare control, reflection, lifetime, heat generation, and ballast selection.

Noise

Noise produced by animals and animal-care activities is inherent in the operation of an animal facility (Pfaff and Stecker 1976). Therefore, noise control should be considered in facility design and operation (Pekrul 1991). Assessment of the potential effects of noise on an animal warrants consideration of the intensity, frequency, rapidity of onset, duration, and vibration potential of the sound and the hearing range, noise-exposure history, and sound-effect susceptibility of the species, stock, or strain.

Separation of human and animal areas minimizes disturbances to both the human and animal occupants of the facility. Noisy animals—such as dogs, swine, goats, and nonhuman primates—should be housed away from quieter animals, such as rodents, rabbits, and cats. Environments should be designed to accommodate animals that make noise, rather than resorting to methods of noise reduction. Exposure to sound louder than 85 dB can have both auditory and nonauditory effects (Fletcher 1976; Peterson 1980), including eosinopenia and increased adrenal weights in rodents (Geber and others 1966; Nayfield and Besch 1981), reduced fertility in rodents (Zondek and Tamari 1964), and increased blood pressure in nonhuman primates (Peterson and others 1981). Many species can hear frequencies of sound that are inaudible to humans (Brown and Pye 1975; Warfield 1973), so the potential effects of equipment and materials that produce noise in the hearing range of nearby animals—such as video display terminals (Sales 1991) should be carefully considered. To the greatest extent possible, activities that might be noisy should be conducted in rooms or areas separate from those used for animal housing.

Because changes in patterns of sound exposure have different effects on different animals (Armario and others 1985; Clough 1982), personnel should try to minimize the production of unnecessary noise. Excessive and intermittent noise can be minimized by training personnel in alternatives to practices that produce noise and by the use of cushioned casters and bumpers on carts, trucks, and racks. Radios, alarms, and other sound generators should not be used in animal rooms unless they are parts of an approved protocol or an enrichment program.

BEHAVIORAL MANAGEMENT

Structural Environment

The structural environment consists of components of the primary enclosure—cage furniture, equipment for environmental enrichment, objects for manipulation by the animals, and cage complexities. Depending on the animal species and use, the structural environment should include resting boards, shelves or perches, toys, foraging devices, nesting materials, tunnels, swings, or other objects that increase opportunities for the expression of species-typical postures and activities and enhance the animals' well-being. Much has been learned in recent years about the natural history and environmental needs of many animals, but continuing research into those environments that enhance the well-being of research animals is encouraged. Selected publications that describe enrichment strategies for common laboratory animal species are listed in Appendix A and in bibliographies prepared by the Animal Welfare Information Center (AWIC 1992; NRC In press).

Social Environment

Consideration should be given to an animal's social needs. The social environment usually involves physical contact and communication among members of the same species (conspecifics), although it can include noncontact communication among individuals through visual, auditory, and olfactory signals. When it is appropriate and compatible with the protocol, social animals should be housed in physical contact with conspecifics. For example, grouping of social primates or canids is often beneficial to them if groups comprise compatible individuals. Appropriate social interactions among conspecifics are essential for normal development in many species. A social companion might buffer the effects of a stressful situation (Gust and others 1994), reduce behavioral abnormality (Reinhardt and others 1988, 1989), increase opportunities for exercise (Whary and others 1993), and expand species-typical behavior and cognitive stimulation. Such factors as population density, ability to disperse, initial familiarity among animals, and social rank should be evaluated when animals are being grouped (Borer and others 1988; Diamond and others 1987; Drickamer 1977; Harvey and Chevins 1987; Ortiz and others 1985; Vandenbergh 1986, 1989). In selecting a suitable social environment, attention should be given to whether the animals are naturally territorial or communal and whether they should be housed singly, in pairs, or in groups. An understanding of species-typical natural social behavior will facilitate successful social housing.

However, not all members of a social species can or should be maintained socially; experimental, health, and behavioral reasons might preclude a successful outcome of this kind of housing. Social housing can increase the likelihood of animal wounds due to fighting (Bayne and others 1995), increase susceptibility to such metabolic disorders as atherosclerosis (Kaplan and others 1982), and alter behavior and physiologic functions (Bernstein 1964; Bernstein and others 1974a,b). In addition, differences between sexes in compatibility have been observed in various species (Crockett and others 1994; Grant and Macintosh 1963; Vandenbergh 1971; vom Saal 1984). These risks of social housing are greatly reduced if the animals are socially compatible and the social unit is stable.

It is desirable that social animals be housed in groups; however, when they must be housed alone, other forms of enrichment should be provided to compensate for the absence of other animals, such as safe and positive interaction with the care staff and enrichment of the structural environment.

Activity

Animal activity typically implies motor activity but also includes cognitive activity and social interaction. Animals maintained in a laboratory environment might have a more-restricted activity profile than those in a free-ranging state. An animal's motor activity, including use of the vertical dimension, should be considered in evaluation of suitable housing or assessment of the appropriateness of the quantity or quality of an activity displayed by an animal. Forced activity for reasons other than attempts to meet therapeutic or approved protocol objectives should be avoided. In most species, physical activity that is repetitive, is non-goal-oriented, and excludes other behavior is considered undesirable (AWIC 1992; Bayne 1991; NRC In press; see also Appendix A, "Enrichment").

Animals should have opportunities to exhibit species-typical activity patterns. Dogs, cats, and many other domesticated animals benefit from positive human interaction (Rollin 1990). Dogs can be given opportunities for activity by being walked on a leash, having access to a run, or being moved into another area (such as a room, larger cage, or outdoor pen) for social contact, play, or exploration. Cages are often used for short-term housing of dogs for veterinary care and some research purposes, but pens, runs, and other out-of-cage areas provide more space for movement, and their use is encouraged (Wolff and Rupert 1991). Loafing areas, exercise lots, and pastures are suitable for large farm animals, such as sheep, horses, and cattle.

HUSBANDRY

Food

Animals should be fed palatable, noncontaminated, and nutritionally adequate food daily or according to their particular requirements unless the protocol in which they are being used requires otherwise. Subcommittees of the National Research Council Committee on Animal Nutrition have prepared comprehensive treatments of the nutrient requirements of laboratory animals (NRC 1977, 1978, 1981a,b, 1982, 1983, 1984, 1985a,b, 1986, 1988, 1989a,b, 1994, 1995). Their publications consider issues of quality assurance, freedom from chemical or microbial contaminants and presence of natural toxicants in feedstuffs, bioavailability of nutrients in feeds, and palatability.

Animal-colony managers should be judicious in purchasing, transporting, storing, and handling food to minimize the introduction of diseases, parasites, potential disease vectors (e.g., insects and other vermin), and chemical contaminants into animal colonies. Purchasers are encouraged to consider manufacturers' and suppliers' procedures and practices for protecting and ensuring diet quality (e.g., storage, vermin-control, and handling procedures). Institutions should urge feed vendors to provide data from feed analysis for critical nutrients periodically. The date of manufacture and other factors that affect shelf-life of food should be known by the user. Stale food or food transported and stored inappropriately can become deficient in nutrients. Careful attention should be paid to quantities received in each shipment, and stock should be rotated so that the oldest food is used first.

Areas in which diets and diet ingredients are processed or stored should be kept clean and enclosed to prevent entry of pests. Food should be stored off the floor on pallets, racks, or carts. Unused, opened bags of food should be stored in vermin-proof containers to minimize contamination and to avoid potential spread of disease agents. Exposure to temperatures above 21°C (70°F), extremes in relative humidity, unsanitary conditions, light, oxygen, and insects and other vermin hasten the deterioration of food. Precautions should be taken if perishable items—such as meats, fruits, and vegetables—are fed, because storage conditions are potential sources of contamination and can lead to variation in food quality. Contaminants in food can have dramatic effects on biochemical and physiologic processes, even if the contaminants are present in concentrations too low to cause clinical signs of toxicity. For example, some contaminants induce the synthesis of hepatic enzymes that can alter an animal's response to drugs (Ames and others 1993; Newberne 1975). Some experimental protocols might require the use of pretested animal diets in which both biologic and nonbiologic contaminants are identified and their concentrations documented.

Most natural-ingredient, dry laboratory-animal diets that contain preservatives and are stored properly can be used up to about 6 months after manufacture. Vitamin C in manufactured feeds, however, generally has a shelf-life of only 3 months. The use of stabilized forms of vitamin C can extend the shelf-life of feed. If a diet containing outdated vitamin C is to be fed to animals that require dietary vitamin C, it is necessary to provide an appropriate vitamin C supplement. Refrigeration preserves nutritional quality and lengthens shelf-life, but food-storage time should be reduced to the lowest practical period and the recommendations of manufacturers should be considered. Purified and chemically defined diets are often less stable than natural-ingredient diets, and their shelf-life is usually less than 6 months (Fullerton and others 1982); these diets should be stored at 4°C (39°F) or lower.

Autoclavable diets require adjustments in nutrient concentrations, kinds of ingredients, and methods of preparation to withstand degradation during sterilization (Wostman 1975). The date of sterilization should be recorded and the diet used quickly. Irradiated diets might be considered as an alternative to autoclaved diets.

Feeders should be designed and placed to allow easy access to food and to minimize contamination with urine and feces. When animals are housed in groups, there should be enough space and enough feeding points to minimize competition for food and ensure access to food for all animals, especially if feed is restricted as part of the protocol or management routine. Food-storage containers should not be transferred between areas that pose different risks of contamination, and they should be cleaned and sanitized regularly.

Moderate restriction of calorie and protein intakes for clinical or husbandry reasons has been shown to increase longevity and decrease obesity, reproduction, and cancer rates in a number of species (Ames and others 1993; Keenan and others 1994). Such restriction can be achieved by decreasing metabolizable energy, protein density, or both in the diet or by controlling ration amount or frequency of feeding. The choice of mechanism for calorie restriction is species-dependent and will affect physiologic adaptations and alter metabolic responses (Leveille and Hanson 1966). Calorie restriction is an accepted practice for long-term housing of some species, such as some rodents and rabbits, and as an adjunct to some clinical and surgical procedures.

In some species (such as nonhuman primates) and on some occasions, varying nutritionally balanced diets and providing "treats," including fresh vegetables, can be appropriate and improve well-being. However, caution should be used in varying diets. A diet should be nutritionally balanced; it is well documented that many animals offered a cafeteria of unbalanced foods do not select a balanced diet and become obese through selection of high-energy, low-protein foods (Moore 1987). Abrupt changes in diet (which are difficult to avoid at weaning) should be minimized because they can lead to digestive and metabolic disturbances; these changes occur in omnivores and carnivores, but herbivores (Eadie and Mann 1970) are especially sensitive.

Water

Ordinarily, animals should have access to potable, uncontaminated drinking water according to their particular requirements. Water quality and the definition of potable water can vary with locality (Homberger and others 1993). Periodic monitoring for pH, hardness, and microbial or chemical contamination might be necessary to ensure that water quality is acceptable, particularly for use in studies in which normal components of water in a given locality can influence the results obtained. Water can be treated or purified to minimize or eliminate contamination when protocols require highly purified water. The selection of water treatments should be carefully considered because many forms of water treatment have the potential to cause physiologic alterations, changes in microflora, or effects on experimental results (Fidler 1977; Hall and others 1980; Hermann and others 1982; Homberger and others 1993). For example, chlorination of the water supply can be useful for some species but toxic to others (such as aquatic species).

Watering devices, such as drinking tubes and automatic waterers, should be checked daily to ensure their proper maintenance, cleanliness, and operation. Animals sometimes have to be trained to use automatic watering devices. It is better to replace water bottles than to refill them, because of the potential for microbiologic cross-contamination; however, if bottles are refilled, care should be taken to replace each bottle on the cage from which it was removed. Animals housed in outdoor facilities might have access to water in addition to that provided in watering devices, such as that available in streams or in puddles after a heavy rainfall. Care should be taken to ensure that such accessory sources of water do not constitute a hazard, but their availability need not routinely be prevented.

Bedding

Animal bedding is a controllable environmental factor that can influence experimental data and animal well-being. The veterinarian or facility manager, in consultation with investigators, should select the most appropriate bedding material. No bedding is ideal for any given species under all management and experimental conditions, and none is ideal for all species (for example, bedding that enables burrowing is encouraged for some species). Several writers (Gibson and others 1987; Jones 1977; Kraft 1980; Thigpen and others 1989; Weichbrod and others 1986) have described desirable characteristics and means of evaluating bedding. Softwood beddings have been used, but the use of untreated softwood shavings and chips is contraindicated for some protocols because they can affect animals' metabolism (Vesell 1967; Vessell and others 1973, 1976). Cedar shavings are not recommended, because they emit aromatic hydrocarbons that induce hepatic microsomal enzymes and cytotoxicity (Torronen and others 1989; Weichbrod and others 1986, 1988) and have been reported to increase the incidence of cancer (Jacobs and Dieter 1978; Vlahakis 1977). Heat treatments applied before bedding materials are used reduce the concentration of aromatic hydrocarbons and might prevent this problem. Manufacturing, monitoring, and storage methods used by vendors should be considered when purchasing bedding products.

Bedding should be transported and stored off the floor on pallets, racks, or carts in a fashion consistent with maintenance of quality and minimization of contamination. During autoclaving, bedding can absorb moisture and as a result lose absorbency and support the growth of microorganisms. Therefore, appropriate drying times and storage conditions should be used.

Bedding should be used in amounts sufficient to keep animals dry between cage changes, and, in the case of small laboratory animals, care should be taken to keep the bedding from coming into contact with the water tube, because such contact could cause leakage of water into the cage.

Sanitation

Sanitation—the maintenance of conditions conducive to health—involves bedding change (as appropriate), cleaning, and disinfection. Cleaning removes excessive amounts of dirt and debris, and disinfection reduces or eliminates unacceptable concentrations of microorganisms.

The frequency and intensity of cleaning and disinfection should depend on what is needed to provide a healthy environment for an animal, in accord with its normal behavior and physiologic characteristics. Methods and frequencies of sanitation will vary with many factors, including the type, physical properties, and size of the enclosure; the type, number, size, age, and reproductive status of the animals; the use and type of bedding materials; temperature and relative humidity; the nature of the materials that create the need for sanitation; the normal physiologic and behavioral characteristics of the animals; and the rate of soiling of the surfaces of the enclosure. Some housing systems or experimental protocols might require specific husbandry techniques, such as aseptic handling or modification in the frequency of bedding change.

Agents designed to mask animal odors should not be used in animal-housing facilities. They cannot substitute for good sanitation practices or for the provision of adequate ventilation, and they expose animals to volatile compounds that might alter basic physiologic and metabolic processes.

Bedding Change

Soiled bedding should be removed and replaced with fresh materials as often as is necessary to keep the animals clean and dry. The frequency is a matter of professional judgment of animal care personnel based on consultation with the investigator and depends on such factors as the number and size of the animals in the primary enclosure, the size of the enclosure, urinary and fecal output, the appearance and wetness of the bedding, and experimental conditions, such as those of surgery or debilitation, that might limit an animal's movement or access to areas of the cage that have not been soiled with urine and feces. There is no absolute minimal frequency of changing bedding, but it typically varies from daily to weekly. In some instances, frequent bedding changes are contraindicated, such as during some portions of the prepartum or postpartum period, when phero-mones are essential for successful reproduction, or when research objectives do not permit changing the bedding.

Cleaning and Disinfection of Primary Enclosures

For pens or runs, frequent flushing with water and periodic use of detergents or disinfectants are usually appropriate to maintain sufficiently clean surfaces. If animal waste is to be removed by flushing, this will need to be done at least once a day. Animals should be kept dry during such flushing. The timing of pen or run cleaning should take into account normal behavioral and physiologic processes of the animals; for example, the gastrocolic reflex in meal-fed animals results in defecation shortly after food consumption.

The frequency of sanitation of cages, cage racks, and associated equipment, such as feeders and watering devices, is governed to some extent by the types of caging and husbandry practices used, including the use of regularly changed contact or noncontact bedding, regular flushing of suspended catch pans, and the use of wire-bottom or perforated-bottom cages. In general, enclosures and accessories, such as tops, should be sanitized at least once every 2 weeks. Solid-bottom caging, bottles, and sipper tubes usually require sanitation at least once a week. Some types of cages and racking might require less-frequent cleaning or disinfection; these might include large cages with very low animal density and frequent bedding changes, cages that house animals in gnotobiotic conditions with frequent bedding changes, individually ventilated cages, and cages used for special circumstances. Some circumstances, such as microisolator housing or more densely populated enclosures, might require more frequent sanitation.

Rabbits and some rodents, such as guinea pigs and hamsters, produce urine with high concentrations of proteins and minerals. Minerals and organic compounds in the urine from these animals often adhere to cage surfaces and necessitate treatment with acid solutions before washing.

Primary enclosures can be disinfected with chemicals, hot water, or a combination of both. Washing times and conditions should be sufficient to kill vegetative forms of common bacteria and other organisms that are presumed to be controllable by the sanitation program. When hot water is used alone, it is the combined effect of the temperature and the length of time that a given temperature (cumulative heat factor) is applied to the surface of the item that disinfects. The same cumulative heat factor can be obtained by exposing organisms to very high temperatures for short periods or exposing them to lower temperatures for longer periods (Wardrip and others 1994). Effective disinfection can be achieved with wash and rinse water at 143-180°F or more. The traditional 82.2°C (180°F) temperature requirement for rinse water refers to the water in the tank or in the sprayer manifold. Detergents and chemical disinfectants enhance the effectiveness of hot water but should be thoroughly rinsed from surfaces before reuse of the equipment.

Washing and disinfection of cages and equipment by hand with hot water and detergents or disinfectants can be effective but require attention to detail. It is particularly important to ensure that surfaces are rinsed free of residual chemicals and that personnel have appropriate equipment to protect themselves from exposure to hot water or chemical agents used in the process.

Water bottles, sipper tubes, stoppers, feeders, and other small pieces of equipment should be washed with detergents, hot water, and, where appropriate, chemical agents to destroy microorganisms.

If automatic watering systems are used, some mechanism to ensure that microorganisms and debris do not build up in the watering devices is recommended. The mechanism can be periodic flushing with large volumes of water or appropriate chemical agents followed by a thorough rinsing. Constant-recirculation loops that use properly maintained filters, ultraviolet lights, or other devices to sterilize recirculated water are also effective.

Conventional methods of cleaning and disinfection are adequate for most animal-care equipment. However, if pathogenic microorganisms are present or if animals with highly defined microbiologic flora or compromised immune systems are maintained, it might be necessary to sterilize caging and associated equipment after cleaning and disinfection. Sterilizers should be regularly calibrated and monitored to ensure their safety and effectiveness.

Cleaning and Disinfection of Secondary Enclosures

All components of the animal facility, including animal rooms and support spaces (such as storage areas, cage-washing facilities, corridors, and procedure rooms) should be cleaned regularly and disinfected as appropriate to the circumstances and at a frequency based on the use of the area and the nature of likely contamination.

Cleaning utensils should be assigned to specific areas and should not be transported between areas that pose different risks of contamination. Cleaning utensils themselves should be cleaned regularly and should be constructed of materials that resist corrosion. Worn items should be replaced regularly. The utensils should be stored in a neat, organized fashion that facilitates drying and minimizes contamination.

Assessing the Effectiveness of Sanitation

Monitoring of sanitation practices should be appropriate to the process and materials being cleaned; it can include visual inspection of the materials, monitoring of water temperatures, or microbiologic monitoring. The intensity of animal odors, particularly that of ammonia, should not be used as the sole means of assessing the effectiveness of the sanitation program. A decision to alter the frequency of cage-bedding changes or cage-washing should be based on such factors as the concentration of ammonia, the appearance of the cage, the condition of the bedding, and the number and size of animals housed in the cage.

Waste Disposal

Conventional, biologic, and hazardous waste should be removed and disposed of regularly and safely (NSC 1979). There are several options for effective waste disposal. Contracts with licensed commercial waste-disposal firms usually provide some assurance of regulatory compliance and safety. On-site incineration should comply with all federal, state, and local regulations.

Adequate numbers of properly labeled waste receptacles should be strategically placed throughout the facility. Waste containers should be leakproof and equipped with tight-fitting lids. It is good practice to use disposable liners and to wash containers and implements regularly. There should be a dedicated waste-storage area that can be kept free of insects and other vermin. If cold storage is used to hold material before disposal, a properly labeled, dedicated refrigerator, freezer, or cold room should be used.

Hazardous wastes must be rendered safe by sterilization, containment, or other appropriate means before being removed from the facility (US EPA 1986). Radioactive wastes should be maintained in properly labeled containers. Their disposal should be closely coordinated with radiation-safety specialists in accord with federal and state regulations. The federal government and most states and municipalities have regulations controlling disposal of hazardous wastes. Compliance with regulations concerning hazardous-agent use (Chapter 1) and disposal is an institutional responsibility.

Infectious animal carcasses can be incinerated on site or collected by a licensed contractor. Procedures for on-site packaging, labeling, transportation, and storage of these wastes should be integrated into occupational health and safety policies.

Hazardous wastes that are toxic, carcinogenic, flammable, corrosive, reactive, or otherwise unstable should be placed in properly labeled containers and disposed of as recommended by occupational health and safety specialists. In some circumstances, these wastes can be consolidated or blended.

Pest Control

Programs designed to prevent, control, or eliminate the presence of or infestation by pests are essential in an animal environment. A regularly scheduled and documented program of control and monitoring should be implemented. The ideal program prevents the entry of vermin into and eliminates harborage from the facility. For animals in outdoor facilities, consideration should also be given to eliminating or minimizing the potential risk associated with pests and predators. Pesticides can induce toxic effects on research animals and interfere with experimental procedures (Ohio Cooperative Extension Service 1987a,b), and they should be used in animal areas only when necessary. Investigators whose animals might be exposed to pesticides should be consulted before pesticides are used. Use of pesticides should be recorded and coordinated with the animal-care management staff and be in compliance with federal, state, or local regulations. Whenever possible, nontoxic means of pest control, such as insect growth regulators (Donahue and others 1989; Garg and Donahue 1989; King and Bennett 1989) and nontoxic substances (for example, amorphous silica gel), should be used. If traps are used, methods should be humane; traps used to catch pests alive require frequent observation and humane euthanasia after capture.

Emergency, Weekend, and Holiday Care

Animals should be cared for by qualified personnel every day, including weekends and holidays, both to safeguard their well-being and to satisfy research requirements. Emergency veterinary care should be available after work hours, on weekends, and on holidays.

In the event of an emergency, institutional security personnel and fire or police officials should be able to reach people responsible for the animals. That can be enhanced by prominently posting emergency procedures, names, or telephone numbers in animal facilities or by placing them in the security department or telephone center. Emergency procedures for handling special facilities or operations should be prominently posted.

A disaster plan that takes into account both personnel and animals should be prepared as part of the overall safety plan for the animal facility. The colony manager or veterinarian responsible for the animals should be a member of the appropriate safety committee at the institution. He or she should be an "official responder" within the institution and should participate in the response to a disaster (Casper 1991).

POPULATION MANAGEMENT

Identification and Records

Means of animal identification include room, rack, pen, stall, and cage cards with written or bar-coded information; collars, bands, plates, and tabs; colored stains; ear notches and tags; tattoos; subcutaneous transponders; and freeze brands. Toe-clipping, as a method of identification of small rodents, should be used only when no other individual identification method is feasible and should be performed only on altricial neonates. Identification cards should include the source of the animal, the strain or stock, names and locations of the responsible investigators, pertinent dates, and protocol number, when applicable. Animal records are useful and can vary in type, ranging from limited information on identification cards to detailed computerized records for individual animals.

Clinical records for individual animals can also be valuable, especially for dogs, cats, nonhuman primates, and farm animals. They should include pertinent clinical and diagnostic information, date of inoculations, history of surgical procedures and postoperative care, and information on experimental use. Basic demographic information and clinical histories enhance the value of individual animals for both breeding and research and should be readily accessible to investigators, veterinary staff, and animal-care staff. Records of rearing histories, mating histories, and behavioral profiles are useful for the management of many species, especially nonhuman primates (NRC 1979a).

Records containing basic descriptive information are essential for management of colonies of large long-lived animals and should be maintained for each animal (Dyke 1993; NRC 1979a). These records often include species, animal identifier, sire identifier, dam identifier, sex, birth or acquisition date, source, exit date, and final disposition. Such animal records are essential for genetic management and historical assessments of colonies. Relevant recorded information should be provided when animals are transferred between institutions.

Genetics and Nomenclature

Genetic characteristics are important in regard to the selection and management of animals for use in breeding colonies and in biomedical research (see Appendix A). Pedigree information allows appropriate selection of breeding pairs and of experimental animals that are unrelated or of known relatedness.

Outbred animals are widely used in biomedical research. Founding populations should be large enough to ensure the long-term heterogeneity of breeding colonies. To facilitate direct comparison of research data derived from outbred animals, genetic-management techniques should be used to maintain genetic variability and equalize founder representations (for example, Lacy 1989; Poiley 1960; Williams-Blangero 1991). Genetic variability can be monitored with computer simulations, biochemical markers, DNA markers, immunologic markers, or quantitative genetic analyses of physiologic variables (MacCluer and others 1986; Williams-Blangero 1993).

Inbred strains of various species, especially rodents, have been developed to address specific research needs (Festing 1979; Gill 1980). The homozygosity of these animals enhances the reproducibility and comparability of some experimental data. It is important to monitor inbred animals periodically for genetic homozygosity (Festing 1982; Hedrich 1990). Several methods of monitoring have been developed that use immunologic, biochemical, and molecular techniques (Cramer 1983; Groen 1977; Hoffman and others 1980; Russell and others 1993). Appropriate management systems (Green 1981; Kempthorne 1957) should be designed to minimize genetic contamination resulting from mutation and mismating.

Transgenic animals have at least one transferred gene whose site of integration and number of integrated copies might or might not have been controlled. Integrated genes can interact with background genes and environmental factors, in part as a function of site of integration, so each transgenic animal can be considered a unique resource. Care should be taken to preserve such resources through standard genetic-management procedures, including maintenance of detailed pedigree records and genetic monitoring to verify the presence and zygosity of transgenes. Cryopreservation of fertilized embryos, ova, or spermatozoa should also be considered to safeguard against alterations in transgenes over time or accidental loss of the colony.

Accurate recording, with standardized nomenclature where it is available, of both the strain and substrain or of the genetic background of animals used in a research project is important (NRC 1979b). Several publications provide rules developed by international committees for standardized nomenclature of outbred rodents and rabbits (Festing and others 1972), inbred rats (Festing and Staats 1973; Gill 1984; NRC 1992a), inbred mice (International Committee on Standardized Genetic Nomenclature for Mice 1981a,b,c), and transgenic animals (NRC 1992b).

REFERENCES

Ames, B. N., M. K. Shigenaga, and T. M. Hagen. 1993. Review: Oxidants, antioxidants, and the degenerative diseases of aging. Proc. Natl. Acad. Sci. USA 90:7915-7922.

Anzaldo, A. J., P. C. Harrison, G. L. Riskowski, L. A. Sebek, R-G. Maghirang, and H. W. Gonyou. 1994. Increasing welfare of laboratory rats with the help of spatially enhanced cages. AWIC Newsl. 5(3):1-2,5.

Armario, A., J. M. Castellanos, and J. Balasch. 1985. Chronic noise stress and insulin secretion in male rats. Physiol. Behav. 34:359-361.

ASHRAE (American Society of Heating, Refrigeration, and Air Conditioning Engineers, Inc.). 1992. Chapter 25: Air Cleaners for Particulate Contaminants. In 1992 ASHRAE Handbook: HVAC Systems and Equipment, I-P edition. Atlanta: ASHRAE.

ASHRAE (American Society of Heating, Refrigeration, and Air Conditioning Engineers, Inc.). 1993. Chapter 9: Environmental Control for Animals and Plants. In 1993 ASHRAE Handbook: Fundamentals, I-P edition. Atlanta: ASHRAE.

AWIC (Animal Welfare Information Center). 1992. Environmental enrichment information resources for nonhuman primates: 1987-1992. National Agricultural Library, US Department of Agriculture; National Library of Medicine, National Institutes of Health; Primate Information Center, University of Washington.

Bayne, K. 1991. Providing environmental enrichment to captive primates. Compendium on Cont. Educ. for the Practicing Vet. 13(11):1689-1695.

Bayne, K., M. Haines, S. Dexter, D. Woodman, and C. Evans. 1995. Nonhuman primate wounding prevalence: A retrospective analysis. Lab Anim. 24(4):40-43.

Bellhorn, R. W. 1980. Lighting in the animal environment. Lab. Anim. Sci. 30(2, Part II):440-450.

Bernstein, I. S. 1964. The integration of rhesus monkeys introduced to a group. Folia Primatol. 2:50-63.

Bernstein, I. S., T. P. Gordon, and R. M. Rose. 1974a. Aggression and social controls in rhesus monkey (*Macaca mulatta*) groups revealed in group formation studies. Folia Primatol. 21:81-107.

Bernstein, I. S., R. M. Rose, and T. P. Gordon. 1974b. Behavioral and environmental events influencing primate testosterone levels. J. Hum. Evol. 3:517-525.

Besch, E. L. 1980. Environmental quality within animal facilities Lab. Anim. Sci. 30:385-406.

Borer, K. T., A. Pryor, C. A. Conn, R. Bonna, and M. Kielb. 1988. Group housing accelerates growth and induces obesity in adult hamsters. Am. J. Physiol. 255(1, Part 2):R128-133.

Brain, P., and D. Bention. 1979. The interpretation of physiological correlates of differential housing in laboratory rats. Life Sci. 24:99-115.

Brainard, G. C. 1989. Illumination of laboratory animal quarters: Participation of light irradiance and wavelength in the regulation of the neuroendocrine system. Pp. 69-74 in Science and Animals: Addressing Contemporary Issues. Greenbelt, Md.: Scientists Center for Animal Welfare.

Brainard, G. C., M. K. Vaughan, and R. J. Reiter. 1986. Effect of light irradiance and wavelength on the Syrian hamster reproductive system. Endocrinology 119(2):648-654.

Broderson, J. R., J. R. Lindsey, and J. E. Crawford. 1976. The role of environmental ammonia in respiratory mycoplasmosis of rats. Am. J. Path. 85:115-127.

Brown, A. M., and J. D. Pye. 1975. Auditory sensitivity at high frequencies in mammals. Adv. Comp. Physiol. Biochem. 6:1-73.

Casper, J. 1991. Integrating veterinary services into disaster management plans. J. Am. Vet. Med. Assoc. 199(4):444-446.

CFR (Code of Federal Regulations). 1985. Title 9 (Animals and Animal Products), Subchapter A (Animal Welfare). Washington, D.C.: Office of the Federal Register.

Cherry, J. A. 1987. The effect of photoperiod on development of sexual behavior and fertility in golden hamsters. Physiol. Behav. 39(4):521-526.

Clough, G. 1982. Environmental effects on animals used in biomedical research. Biol. Rev. 57:487-523.

Cramer, D. V. 1983. Genetic monitoring techniques in rats. ILAR News 26(4):15-19.

Crockett, C. M., C. L. Bowers, G. P. Sackett, and D. M. Bowden. 1993. Urinary cortisol responses of longtailed macaques to five cage sizes, tethering, sedation, and room change. Am. J. Primatol. 30:55-74.

Crockett, C. M., C. L. Bowers, D. M. Bowden, and G. P. Sackett. 1994. Sex differences in compatibility of pair-housed adult longtailed macaques. Am. J. Primatol. 32:73-94.

Crockett, C. M., C. L. Bowers, M. Shimoji, M. Leu, D. M. Boween, and G. P. Sackett. 1995. Behavioral responses of longtailed macaques to different cage sizes and common laboratory experiences. J. Comp. Psychol. 109(4):368-383.

Diamond, M. C., E. R. Greer, A. York, D. Lewis, T. Barton, and J. Lin. 1987. Rat cortical morphology following crowded-enriched living conditions. Exp. Neurol. 96(2):241-247.

Donahue, W. A., D. N. VanGundy, W. C. Satterfield, and L. G. Coghlan. 1989. Solving a tough problem. Pest Control :46-50.

Drickamer, L. C. 1977. Delay of sexual maturation in female house mice by exposure to grouped females or urine from grouped females. J. Reprod. Fertil. 51:77-81.

Duncan, T. E., and W. K. O'Steen. 1985. The diurnal susceptibility of rat retinal photoreceptors to light-induced damage. Exp. Eye Res. 41(4):497-507.

Dyke, B. 1993. Basic data standards for primate colonies. Am. J. Primatol. 29:125-143.

Eadie, J. M., and S. O. Mann. 1970. Development of the rumen microbial population: High starch diets and instability. Pp. 335-347 in Physiology of Digestion and Metabolism in the Ruminant. Proceedings of the Third International Symposium, A. T. Phillipson, E. F. Annison, D. G. Armstrong, C. C. Balch, R. S. Comline, R. N. Hardy, P. N. Hobson, and R. D. Keynes, eds. Newcastle upon Tyne, England: F.R.S. Oriel Press Limited.

Erkert, H. G., and J. Grober. 1986. Direct modulation of activity and body temperature of owl monkeys (*Aotus lemurinus griseimembra*) by low light intensities. Folia Primatol. 47(4):171-188.

Festing, M. F. W. 1979. Inbred Strains in Biomedical Research. London: MacMillan Press. 483 pp.

Festing, M. F. W. 1982. Genetic contamination of laboratory animal colonies: an increasingly serious problem. ILAR News 25(4):6-10.

Festing, M., and J. Staats. 1973. Standardized nomenclature for inbred strains of rats. Fourth listing. Transplantation 16(3):221-245.

Festing, M. F. W., K. Kondo, R. Loosli, S. M. Poiley, and A. Spiegel. 1972. International standardized nomenclature for outbred stocks of laboratory animals. ICLA Bull. 30:4-17.

Fidler, I. J. 1977. Depression of macrophages in mice drinking hyperchlorinated water. Nature 270:735-736.

Fletcher, J. L. 1976. Influence of noise on animals. Pp. 51-62 in Control of the Animal House Environment. Laboratory Animal Handbooks 7, T. McSheehy, ed. London: Laboratory Animals Ltd.

Flynn, R. J. 1959. Studies on the aetiology of ringtail of rats. Proc. Anim. Care Panel 9:155-160.

Fullerton, P. M., and R. W. Gilliatt. 1967. Pressure neuropathy in the hind foot of the guinea pig. J. Neurol. Neurosurg. Psychiatry 30:18-25.

Fullerton, F. R., D. L. Greenman, and D. C. Kendall. 1982. Effects of storage conditions on nutritional qualities of semipurified (AIN-76) and natural ingredient (NIH-07) diets. J. Nutr. 112(3):567-473.

Gamble, M. R., and G. Clough. 1976. Ammonia build-up in animal boxes and its effect on rat tracheal epithelium. Lab. Anim. (London) 10(2):93-104.

Garg, R. C., and W. A. Donahue. 1989. Pharmacologic profile of methoprene, and insect growth regulator, in cattle, dogs, and cats. J. Am. Vet. Med. Assoc. 194(3):410-412.

Garrard, G., G. A. Harrison, and J. S. Weiner. 1974. Reproduction and survival of mice at 23°C. J. Reprod. Fertil. 37:287-298.

Geber, W. F., T. A. Anderson, and B. Van Dyne. 1966. Physiologic responses of the albino rat to chronic noise stress. Arch. Environ. Health 12:751-754.

Gibson, S. V., C. Besch-Williford, M. F. Raisbeck, J. E. Wagner, and R. M. McLaughlin. 1987. Organophosphate toxicity in rats associated with contaminated bedding. Lab. Anim. 37(6):789-791.

Gill, T. J. 1980. The use of randomly bred and genetically defined animals in biomedical research. Am. J. Pathol. 101(3S):S21-S32.

Gill, T. J., III. 1984. Nomenclature of alloantigenic systems in the rat. ILAR News 27(3):11-12.

Gordon, C. J. 1990. Thermal biology of the laboratory rat. Physiol. Behav. 47:963-991.

Gordon, C. J. 1993. Temperature Regulation in Laboratory Animals. New York: Cambridge University Press.

Grant, E. C., and J. H. Mackintosh. 1963. A comparison of the social postures of some common laboratory rodents. Behavior 21:246-259.

Green, E. L. 1981. Genetics and Probability in Animal Breeding Experiments. New York: Oxford University Press. 271 pp.

Greenman, D. L., P. Bryant, R. L. Kodell, and W. Sheldon. 1982. Influence of cage shelf level on retinal atrophy in mice. Lab. Anim. Sci. 32(4):353-356.

Groen, A. 1977. Identification and genetic monitoring of mouse inbred strains using biomedical polymorphisms. Lab. Anim. (London) II(4):209-214.

Grover-Johnson, N., and P. S. Spencer. 1981. Peripheral nerve abnormalities in aging rats. J. Neuropath. Exp. Neurol. 40(2):155-165.

Gust, D. A., T. P. Gordon, A. R. Bridie, and H. M. McClure. 1994. Effect of a preferred companion in modulating stress in adult female rhesus monkeys. Physiol. Behav. 55(4):681-684.

Hall, J. E., W. J. White, and C. M. Lang. 1980. Acidification of drinking water: Its effects on selected biologic phenomena in male mice. Lab. Anim. Sci. 30:643-651.

Harvey, P. W., and P. F. D. Chevins. 1987. Crowding during pregnancy delays puberty and alters estrous cycles of female offspring in mice. Experientia 43(3):306-308.

Hedrich, H. J. 1990. Genetic Monitoring of Inbred Strains of Rats. New York: Gustav, Fischer Verlag. 539 pp.

Hermann, L. M., W. J. White, and C. M. Lang. 1982. Prolonged exposure to acid, chlorine, or tetracycline in drinking water: Effects on delayed-type hypersensitivity, hemagglutination titers, and reticuloendothelial clearance rates in mice. Lab. Anim. Sci. 32:603-608.

Hoffman, H. A., K. T. Smith, J. S. Crowell, T. Nomura, and T. Tomita. 1980. Genetic quality control of laboratory animals with emphasis on genetic monitoring. Pp. 307-317 in Animal Quality and Models in Biomedical Research, A. Spiegel, S. Erichsen, and H. A. Solleveld, eds. Stuttgart: Gustav Fischer Verlag.

Homberger, F. R., Z. Pataki, and P. E. Thomann. 1993. Control of *Pseudomonas aeruginosa* infection in mice by chlorine treatment of drinking water. Lab. Anim. Sci. 43(6):635-637.

Hughes, H. C., and S. Reynolds. 1995. The use of computational fluid dynamics for modeling air flow design in a kennel facility. Contemp. Topics 34:49-53.

International Committee on Standardized Genetic Nomenclature for Mice. 1981a. Rules and guidelines for gene nomenclature. Pp. 1-7 in Genetic Variants and Strains of the Laboratory Mouse, M. C. Green, ed. Stuttgart: Gustav Fischer Verlag.

International Committee on Standardized Genetic Nomenclature for Mice. 1981b. Rules for the nomenclature of chromosome abnormalities. Pp. 314-316 in Genetic Variants and Strains of the Laboratory Mouse, M. C. Green, ed. Stuttgart: Gustav Fischer Verlag.

International Committee on Standardized Genetic Nomenclature for Mice. 1981c. Rules for the nomenclature of inbred strains. Pp. 368-372 in Genetic Variants and Strains of the Laboratory Mouse, M. C. Green, ed. Stuttgart: Gustav Fischer Verlag.

Jacobs, B. B., and D. K. Dieter. 1978. Spontaneous hepatomas in mice inbred from Ha:ICR swiss stock: Effects of sex, cedar shavings in bedding, and immunization with fetal liver or hepatoma cells. J. Natl. Cancer Inst. 61(6):1531-1534.

Jones, D. M. 1977. The occurrence of dieldrin in sawdust used as bedding material. Lab. Anim. 11:137.

Kaplan, J. R., S. B. Manuck, T. B. Clarkson, F. M. Lusso, and D. M. Taub. 1982. Social status, environment, and atherosclerosis in cynomolgus monkeys. Arteriosclerosis 2(5):359-368.

Kaufman, J. E. 1984. IES Lighting Handbook Reference Volume. New York: Illuminating Engineering Society.

Kaufman, J. E.. 1987. IES Lighting Handbook Application Volume. New York: Illuminating Engineering Society.

Keenan, K. P., P. F. Smith, and K. A. Soper. 1994. Effect of dietary (caloric) restriction on aging, survival, pathobiology and toxicology. Pp. 609-628 in Pathobiology of the Aging Rat, vol. 2, W. Notter, D. L. Dungworth, and C. C. Capen, eds. International Life Sciences Institute.

Kempthorne, O. 1957. An Introduction to Genetic Statistics. New York: John Wiley and Sons.

King, J. E., and G. W. Bennett. 1989. Comparative activity of fenoxycarb and hydroprene in sterilizing the German cockroach (Dictyoptera: Blattellidae). J. Econ. Entomol. 82(3):833-838.

Kraft, L. M. 1980. The manufacture, shipping and receiving, and quality control of rodent bedding materials. Lab. Anim. Sci. 30(2):366-376.

Lacy, R. C. 1989. Analysis of founder representation in pedigrees: Founder equivalents and founder genome equivalents. Zoo Biology 8:111-123.

Lanum, J. 1979. The damaging effects of light on the retina: Empirical findings, theoretical and practical implications. Surv. Ophthalmol. 22:221-249.

Larson, R. E., and R. O. Hegg. 1976. Feedlot and Ranch Equipment for Beef Cattle. Farmers' Bulletin No. 1584. Washington, D.C.: Agricultural Research Service, U.S. Department of Agriculture. 20 pp.

Leveille, G. A., and R. W. Hanson. 1966. Adaptive changes in enzyme activity and metabolic pathways in adipose tissue from meal-fed rats. J. Lipid Res. 7:46.

MacCluer, J. W., J. L. VandeBerg, B. Read, and O. A. Ryder. 1986. Pedigree analysis by computer simulation. Zoo Biology 5:147-160.

Midwest Plan Service. 1987. Structures and Environment Handbook. 11th ed. rev. Ames: Midwest Plan Service, Iowa State University.

Moore, B. J. 1987. The California diet: An inappropriate tool for studies of thermogenesis. J. Nutr. 117(2):227-231.

Murakami, H. 1971. Differences between internal and external environments of the mouse cage. Lab. Anim. Sci. 21(5):680-684.

NASA (National Aeronautics and Space Administration). 1988. Summary of conclusions reached in workshop and recommendations for lighting animal housing modules used in microgravity related projects. Pp. 5-8 in Lighting Requirements in Microgravity: Rodents and Nonhuman Primates. NASA Technical Memorandum 101077, D. C. Holley, C. M. Winget, and H. A. Leon, eds. Moffett Field, Calif.: Ames Research Center. 273 pp.

Nayfield, K. C., and E. L. Besch. 1981. Comparative responses of rabbits and rats to elevated noise. Lab. Anim. Sci. 31(4):386-390.

Newberne, P. M. 1975. Influence on pharmacological experiments of chemicals and other factors in diets of laboratory animals. Fed. Proc. 34(2):209-218.

Newbold, J. A., L. T. Chapin, S. A. Zinn, and H. A. Tucker. 1991. Effects of photoperiod on mammary development and concentration of hormones in serum of pregnant dairy heifers. J. Dairy Sci. 74(1):100-108.

NRC (National Research Council). 1977. Nutrient Requirements of Rabbits. A report of the Committee on Animal Nutrition. Washington, D.C.: National Academy Press.

NRC (National Research Council). 1978. Nutrient Requirements of Nonhuman Primates. A report of the Committee on Animal Nutrition. Washington, D.C.: National Academy Press.

NRC (National Research Council). 1979a. Laboratory Animal Records. A report of the Committee on Laboratory Animal Records. Washington, D. C.: National Academy Press.

NRC (National Research Council). 1979b. Laboratory animal management: Genetics. A report of the Institute of Laboratory Animal Resources. ILAR News 23(1):A1-A16.

NRC (National Research Council). 1981a. Nutrient Requirements of Cold Water Fishes. A report of the Committee on Animal Nutrition. Washington, D.C.: National Academy Press.

NRC (National Research Council). 1981b. Nutrient Requirements of Goats. A report of the Committee on Animal Nutrition. Washington, D.C.: National Academy Press.

NRC (National Research Council). 1982. Nutrient Requirements of Mink and Foxes. A report of the Committee on Animal Nutrition. Washington, D.C.: National Academy Press.

NRC (National Research Council). 1983. Nutrient Requirements of Warm Water Fishes and Shellfishes. A report of the Committee on Animal Nutrition. Washington, D.C.: National Academy Press.

NRC (National Research Council). 1984. Nutrient Requirements of Beef Cattle. A report of the Committee on Animal Nutrition. Washington, D.C.: National Academy Press.

NRC (National Research Council). 1985a. Nutrient Requirements of Dogs. A report of the Committee on Animal Nutrition. Washington, D.C.: National Academy Press.

NRC (National Research Council). 1985b. Nutrient Requirements of Sheep. A report of the Committee on Animal Nutrition. Washington, D.C.: National Academy Press.

NRC (National Research Council). 1986. Nutrient Requirements of Cats. A report of the Committee on Animal Nutrition. Washington, D.C.: National Academy Press.

NRC (National Research Council). 1988. Nutrient Requirements of Swine. A report of the Committee on Animal Nutrition. Washington, D.C.: National Academy Press.

NRC (National Research Council). 1989a. Nutrient Requirements of Horses. A report of the Committee on Animal Nutrition. Washington, D.C.: National Academy Press.

NRC (National Research Council). 1989b. Nutrient Requirements of Dairy Cattle. A report of the Committee on Animal Nutrition. Washington, D.C.: National Academy Press.

NRC (National Research Council). 1992a. Definition, nomenclature, and conservation of rat strains. A report of the Institute of Laboratory Animal Resources Committee on Rat Nomenclature. ILAR News 34(4):S1-S26.

NRC (National Research Council). 1992b. Standardized nomenclature for transgenic animals. A report of the Institute of Laboratory Animal Resources Committee on Transgenic Nomenclature. ILAR News 34(4):45-52.

NRC (National Research Council). 1994. Nutrient Requirements of Poultry. A report of the Committee on Animal Nutrition. Washington, D.C.: National Academy Press.

NRC (National Research Council). 1995. Nutrient Requirements of Laboratory Animals. A report of the Committee on Animal Nutrition. Washington, D.C.: National Academy Press.

NRC (National Research Council). In press. Psychological Well-being of Nonhuman Primates. A report of the Institute of Laboratory Animal Resources Committee on Well-being of Nonhuman Primates. Washington, D.C.: National Academy Press.

NSC (National Safety Council). 1979. Disposal of potentially contaminated animal wastes. Data sheet 1-679-79. Chicago: National Safety Council.

Ohio Cooperative Extension Service. 1987a. Pesticides for Poultry and Poultry Buildings. Columbus, Ohio: Ohio State University.

Ohio Cooperative Extension Service. 1987b. Pesticides for Livestock and Farm Buildings. Columbus, Ohio: Ohio State University.

Ortiz, R., A. Armario, J. M. Castellanos, and J. Balasch. 1985. Post-weaning crowding induces corticoadrenal hyperactivity in male mice. Physiol. Behav. 34(6):857-860.

Ortman, J. A., J. Sahenk, and J. R. Mendell. 1983. The experimental production of Renault bodies. J. Neurol. Sci. 62:233-241.

O'Steen, W. K. 1980. Hormonal influences in retinal photodamage, Pp. 29-49 in The Effects of Constant Light on Visual Processes, T. P. Williams and B. N. Baker, eds. New York: Plenum Press.

Pekrul, D. 1991. Noise control. Pp. 166-173 in Handbook of Facilities Planning. Vol. 2: Laboratory Animal Facilities, T. Ruys, ed. New York: Van Nostrand Reinhold. 422 pp.

Pennycuik, P. R. 1967. A comparison of the effects of a range of high environmental temperatures and of two different periods of acclimatization on the reproductive performances of male and female mice. Aust. J. Exp. Biol. Med. Sci. 45:527-532.

Peterson, E. A. 1980. Noise and laboratory animals. Lab. Anim. Sci. 30(2, Part II):422-439.

Peterson, E. A., J. S. Augenstein, D. C. Tanis, and D. G. Augenstein. 1981. Noise raises blood pressure without impairing auditory sensitivity. Science 211:1450-1452.

Pfaff, J., and M. Stecker. 1976. Loudness levels and frequency content of noise in the animal house. Lab. Anim. (London) 10(2):111-117.

Poiley, S. M. 1960. A systematic method of breeder rotation for non-inbred laboratory animal colonies. Proc. Anim. Care Panel 10(4):159-166.

Reinhardt, V. D., D. Houser, S. Eisele, D. Cowley, and R. Vertein. 1988. Behavioral responses of unrelated rhesus monkey females paired for the purpose of environmental enrichment. Am. J. Primatol. 14:135-140.

Reinhardt, V. 1989. Behavioral responses of unrelated adult male rhesus monkeys familiarized and paired for the purpose of environmental enrichment. Am. J. Primatol. 17:243-248.

Reynolds, S. D., and H. C. Hughes. 1994. Design and optimization of air flow patterns. Lab Anim. 23:46-49.

Rollin, B. E. 1990. Ethics and research animals: theory and practice. Pp. 19-36 in The Experimental Animal in Biomedical Research. Vol. I: A Survey of Scientific and Ethical Issues for Investigators. B. Rollin and M. Kesel, eds. Boca Raton, Fla.: CRC Press.

Russell, R. J., M. F. W. Festing, A. A. Deeny, and A. G. Peters. 1993. DNA fingerprinting for genetic monitoring of inbred laboratory rats and mice. Lab. Anim. Sci. 43:460-465.

Sales, G. D. 1991. The effect of 22 kHz calls and artificial 38 kHz signals on activity in rats. Behav. Processes 24:83-93.

Saltarelli, D. G., and C. P. Coppola. 1979. Influence of visible light on organ weights of mice. Lab. Anim. Sci. 29(3):319-322.

Schoeb, T. R., M. K. Davidson, and J. R. Lindsey. 1982. Intracage ammonia promotes growth of mycoplasma pulmonis in the respiratory tract of rats. Infect. Immun. 38:212-217.

Semple-Rowland, S. L., and W. W. Dawson. 1987. Retinal cyclic light damage threshold for albino rats. Lab. Anim. Sci. 37(3)289-298.

Serrano, L. J. 1971. Carbon dioxide and ammonia in mouse cages: Effect of cage covers, population and activity. Lab. Anim. Sci. 21(1):75-85.

Stoskopf, M. K. 1983. The physiological effects of psychological stress. Zoo Biology 2:179-190.

Stricklin, W. R. 1995. Space as environmental enrichment. Lab. Anim. 24(4):24-29.

Thigpen, J. E., E. H. Lebetkin, M. L. Dawes, J. L. Clark, C. L. Langley, H. L. Amy, and D. Crawford. 1989. A standard procedure for measuring rodent bedding particle size and dust content. Lab. Anim. Sci. 39(1):60-62.

Torronen, R., K. Pelkonen, and S. Karenlampi. 1989. Enzyme-inducing and cytotoxic effects of wood-based materials used as bedding for laboratory animals. Comparison by a cell culture study. Life Sci. 45:559-565.

Tucker, H. A., D. Petitclerc, and S. A. Zinn. 1984. The influence of photoperiod on body weight gain, body composition, nutrient intake and hormone secretion. J. Anim. Sci. 59(6):1610-1620.

US EPA (U.S. Environmental Protection Agency). 1986. EPA guide for infectious waste management. Washington D.C.: U.S. Environmental Protection Agency; Publication no. EPA/530-5W-86-014.

Vandenbergh, J. G. 1971. The effects of gonadal hormones on the aggressive behavior of adult golden hamsters. Anim. Behav. 19:585-590.

Vandenbergh, J. G. 1986. The suppression of ovarian function by chemosignals. Pp. 423-432 in Chemical Signals in Vertebrates 4, D. Duvall, D. Muller-Schwarze, and R. M. Silverstein, eds. New York: Plenum Publishing.

Vandenbergh, J. G. 1989. Coordination of social signals and ovarian function during sexual development. J. Anim. Sci. 67:1841-1847.

Vesell, E. S. 1967. Induction of drug-metabolizing enzymes in liver microsomes of mice and rats by softwood bedding. Science 157:1057-1058.

Vesell, E. S., C. M. Lang, W. J. White, G. T. Passananti, and S. L. Tripp. 1973. Hepatic drug metabolism in rats: Impairment in a dirty environment. Science 179:896-897.

Vesell, E. S., C. M. Lang, W. J. White, G. T. Passananti, R. N. Hill, T. L. Clemens, D. L. Liu, and W. D. Johnson. 1976. Environmental and genetic factors affecting response of laboratory animals to drugs. Fed. Proc. 35:1125-1132.

Vlahakis, G. 1977. Possible carcinogenic effects of cedar shavings in bedding of C3H-A^{vy}fB mice. J. Natl. Cancer Inst. 58(1):149-150.

vom Saal, F. 1984. The intrauterine position phenomenon: Effects on physiology, aggressive behavior and population dynamics in house mice. Pp. 135-179 in Biological Perspectives on Aggression, K. Flannelly, R. Blanchard, and D. Blanchard, eds. Prog. Clin. Biol. Res. Vol. 169 New York: Alan Liss.

Wardrip, C. L., J. E. Artwohl, and B. T. Bennett. 1994. A review of the role of temperature versus time in an effective cage sanitation program. Contemp. Topics 33:66-68.

Warfield, D. 1973. The study of hearing in animals. Pp. 43-143 in Methods of Animal Experimentation, IV, W. Gay, ed. London: Academic Press.

Wax, T. M. 1977. Effects of age, strain, and illumination intensity on activity and self-selection of light-dark schedules in mice. J. Comp. Physiol. Psychol. 91(1):51-62.

Weichbrod, R. H., J. E. Hall, R. C. Simmonds, and C. F. Cisar. 1986. Selecting bedding material. Lab Anim. 15(6):25-9.

Weichbrod, R. H., C. F. Cisar, J. G. Miller, R. C. Simmonds, A. P. Alvares, and T. H. Ueng. 1988. Effects of cage beddings on microsomal oxidative enzymes in rat liver. Lab. Anim. Sci. 38(3):296-8.

Whary, M., R. Peper, G. Borkowski, W. Lawrence, and F. Ferguson. 1993. The effects of group housing on the research use of the laboratory rabbit. Lab. Anim. 27:330-341.

White, W. J. 1990. The effects of cage space and environmental factors. Pp. 29-44 in Guidelines for the Well-being of Rodents in Research, H. N. Guttman, ed. Proceedings from a conference organized by the Scientists Center for Animal Welfare and held December 9, 1989, in Research Triangle Park, North Carolina. Bethesda, Md.: Scientists Center for Animal Welfare.

White, W. J., M. W. Balk, and C. M. Lang. 1989. Use of cage space by guinea pigs. Lab. Anim. (London) 23:208-214.

Williams-Blangero, S. 1991. Recent trends in genetic research on captive and wild nonhuman primate populations. Year. Phys. Anthropol. 34:69-96.

Williams-Blangero, S. 1993. Research-oriented genetic management of nonhuman primate colonies. Lab. Anim. Sci. 43:535-540.

Wolff, A., and Rupert, G. 1991. A practical assessment of a nonhuman primate exercise program. Lab. Anim. 20(2):36-39.

Wostman, B. S. 1975. Nutrition and metabolism of the germfree mammal. World Rev. Nutr. Diet. 22:40-92.

Zondek, B., and I. Tamari. 1964. Effect of audiogenic stimulation on genital function and reproduction. III. Infertility induced by auditory stimuli prior to mating. Acta Endocrinol. 45(Suppl. 90):227-234.

3

Veterinary Medical Care

Veterinary medical care is an essential part of an animal care and use program. Adequate veterinary care consists of effective programs for

- Preventive medicine.
- Surveillance, diagnosis, treatment, and control of disease, including zoonosis control.
- Management of protocol-associated disease, disability, or other sequelae.
- Anesthesia and analgesia.
- Surgery and postsurgical care.
- Assessment of animal well-being.
- Euthanasia.

A veterinary-care program is the responsibility of the attending veterinarian, who is certified (see ACLAM, Appendix B) or has training or experience in laboratory animal science and medicine or in the care of the species being used. Some aspects of the veterinary-care program can be conducted by persons other than a veterinarian, but a mechanism for direct and frequent communication should be established to ensure that timely and accurate information is conveyed to the veterinarian on problems associated with animal health, behavior, and well-being. The veterinarian must provide guidance to investigators and all personnel involved in the care and use of animals to ensure appropriate handling, immobilization, sedation, analgesia, anesthesia, and euthanasia. The attending veterinarian must provide guidance or oversight to surgery programs and oversight of postsurgical care.

ANIMAL PROCUREMENT AND TRANSPORTATION

All animals must be acquired lawfully, and the receiving institution should make reasonable attempts to ensure that all transactions involving animal procurement are conducted in a lawful manner. If dogs and cats are obtained from USDA Class B dealers or pounds, the animals should be inspected to see whether they can be identified, as through the presence of tattoos or subcutaneous transponders. Such identification might indicate that an animal was a pet, and ownership should be verified. Attention should be given to the population status of the taxon under consideration; the threatened or endangered status of species is provided and updated annually by the Fish and Wildlife Service (DOI 50 CFR 17). The use of purpose-bred research animals might be desirable if it is consistent with research, teaching, and testing objectives.

Potential vendors should be evaluated for the quality of animals supplied by them. As a rule, vendors of purpose-bred animals (e.g., USDA Class A dealers) regularly provide information that describes the genetic and pathogen status of their colonies or individual animals. This information is useful for deciding on acceptance or rejection of animals, and similar data should be obtained on animals received by interinstitutional or intrainstitutional transfer (such as transgenic mice).

All transportation of animals, including intrainstitutional transportation, should be planned to minimize transit time and the risk of zoonoses, protect against environmental extremes, avoid overcrowding, provide food and water when indicated, and protect against physical trauma. Some transportation-related stress is inevitable, but it can be minimized by attention to those factors. Each shipment of animals should be inspected for compliance with procurement specifications and signs of clinical disease and should be quarantined and stabilized according to procedures appropriate for the species and the circumstances. Coordination of ordering and receiving with animal-care personnel is important to ensure that animals are received properly and that appropriate facilities are available for housing.

Several documents provide details on transportation, including the AWRs and the International Air Transport Association Live Animal Regulations (IATA 1995). In addition, import of primates is regulated by the Public Health Service (CFR Title 42) with specific guidelines for tuberculin testing (CDC 1993). There are special requirements for importing and transporting African green, cynomolgus, and rhesus monkeys (FR 1990; CDC 1991).

PREVENTIVE MEDICINE

Disease prevention is an essential component of comprehensive veterinary medical care. Effective preventive-medicine programs enhance the research value of animals by maintaining healthy animals and minimizing nonprotocol sources of variation associated with disease and inapparent infection. These programs consist of various combinations of policies, procedures, and practices related to quarantine and stabilization and the separation of animals by species, source, and health status.

Quarantine, Stabilization, and Separation

Quarantine is the separation of newly received animals from those already in the facility until the health and possibly the microbial status of the newly received animals have been determined. An effective quarantine minimizes the chance for introduction of pathogens into an established colony. The veterinary medical staff should have procedures for evaluating the health and, if appropriate, the pathogen status of newly received animals, and the procedures should reflect acceptable veterinary medical practice and federal and state regulations applicable to zoonoses (Butler and others 1995). Effective quarantine procedures should be used for nonhuman primates to help limit exposure of humans to zoonotic infections. Filoviral and mycobacterial infections in nonhuman primates have recently necessitated specific guidelines for handling nonhuman primates (CDC 1991, 1993). Information from vendors on animal quality should be sufficient to enable a veterinarian to determine the length of quarantine, to define the potential risks to personnel and animals within the colony, to determine whether therapy is required before animals are released from quarantine, and, in the case of rodents, to determine whether cesarean rederivation or embryo transfer is required to free the animals of specific pathogens. Rodents might not require quarantine if data from the vendor or provider are sufficiently current and complete to define the health status of the incoming animals and if the potential for exposure to pathogens during transit is considered. When quarantine is indicated, animals from one shipment should be separated from animals from other shipments (not necessarily from each other) to preclude transfer of infectious agents between groups.

Regardless of the duration of quarantine, newly received animals should be given a period for physiologic, psychologic, and nutritional stabilization before their use. The length of time for stabilization will depend on the type and duration of animal transportation, the species involved, and the intended use of the animals. The need for a stabilization period has been demonstrated in mice, rats, guinea pigs, and goats; it is probably required for other species as well (Drozdowicz and others 1990; Jelinek 1971; Landi and others 1982; Prasad and others 1978; Sanhouri and others 1989; Tuli and others 1995; Wallace 1976).

Physical separation of animals by species is recommended to prevent interspecies disease transmission and to eliminate anxiety and possible physiologic and behavioral changes due to interspecies conflict. Such separation is usually accomplished by housing different species in separate rooms; however, cubicles, laminar-flow units, cages that have filtered air or separate ventilation, and isolators might be suitable alternatives. In some instances, it might be acceptable to house different species in the same room, for example, if two species have a similar pathogen status and are behaviorally compatible. Some species can have subclinical or latent infections that can cause clinical disease if transmitted to another species. A few examples might serve as a guide in determining the need for separate housing by species:

- *Bordetella bronchiseptica* characteristically produces only subclinical infections in rabbits, but severe respiratory disease might occur in guinea pigs (Manning and others 1984).
- As a rule, New World (South American), Old World African, and Old World Asian species of nonhuman primates should be housed in separate rooms. Simian hemorrhagic fever (Palmer and others 1968) and simian immunodeficiency virus (Hirsch and others 1991; Murphey-Corb and others 1986), for example, cause only subclinical infections in African species but induce clinical disease in Asian species.
- Some species should be housed in separate rooms even though they are from the same geographic region. Squirrel monkeys (*Saimiri sciureus*), for example, might be latently infected with *Herpesvirus tamarinus*, which can be transmitted to and cause a fatal epizootic disease in owl monkeys (*Aotus trivirgatus*) (Hunt and Melendez 1966) and some species of marmosets and tamarins (*Saguinus oedipus, S. nigricollis*) (Holmes and others 1964; Melnick and others 1964).

Intraspecies separation might be essential when animals obtained from multiple sites or sources, either commercial or institutional, differ in pathogen status, e.g., sialodacryoadenitis virus in rats, mouse hepatitis virus, *Pasteurella multocida* in rabbits, for *Cercopithecine herpesvirus 1* (formerly *Herpesvirus simiae*) in macaque species, and *Mycoplasma hyopneumoniae* in swine.

Surveillance, Diagnosis, Treatment, and Control of Disease

All animals should be observed for signs of illness, injury, or abnormal behavior by a person trained to recognize such signs. As a rule, this should occur daily, but more-frequent observations might be warranted, such as during postoperative recovery or when animals are ill or have a physical deficit. There might also be situations in which daily observations of each animal is impractical, for example, when animals are housed in large outdoor settings. Professional judgment should be used to ensure that the frequency and character of observation minimize risks to individual animals.

It is imperative that appropriate methods be in place for disease surveillance and diagnosis. Unexpected deaths and signs of illness, distress, or other deviations from normal in animals should be reported promptly to ensure appropriate and timely delivery of veterinary medical care. Animals that show signs of a contagious disease should be isolated from healthy animals in the colony. If an entire room of animals is known or believed to be exposed to an infectious agent (e.g., *Mycobacterium tuberculosis* in nonhuman primates), the group should be kept intact during the process of diagnosis, treatment, and control.

Methods of disease prevention, diagnosis, and therapy should be those currently accepted in veterinary practice. Diagnostic laboratory services facilitate veterinary medical care and can include gross and microscopic pathology, clinical pathology, hematology, microbiology, clinical chemistry, and serology. The choice of medication or therapy should be made by the veterinarian in consultation with the investigator. The selected treatment plan should be therapeutically sound and, when possible, should cause no undesirable experimental variable.

Subclinical microbial, particularly viral, infections (see Appendix A) occur frequently in conventionally maintained rodents but also can occur in facilities designed and maintained for production and use of pathogen-free rodents if a component of the microbial barrier is breached. Examples of infectious agents that can be subclinical but induce profound immunologic changes or alter physiologic, pharmacologic, or toxicologic responses are Sendai virus, Kilham rat virus, mouse hepatitis virus, lymphocytic choriomeningitis virus, and *Mycoplasma pulmonis* (NRC 1991a,b). Scientific objectives of a particular protocol, the consequences of infection within a specific strain of rodent, and the adverse effects that infectious agents might have on other protocols in a facility should determine the characteristics of rodent health-surveillance programs and strategies for keeping rodents free of specific pathogens.

The principal method for detecting viral infections is serologic testing. Other methods of detecting microbial infections, such as bacterial culturing and histopathology and DNA analysis using the polymerase chain reaction (PCR), should be used in combinations that are most suitable for specific requirements of clinical and research programs. Transplantable tumors, hybridomas, cell lines, and other biologic materials can be sources of murine viruses that can contaminate rodents (Nicklas and others 1993). The mouse-antibody-production (MAP), rat-antibody-production (RAP), and hamster-antibody-production (HAP) tests are effective in monitoring for viral contamination of biologic materials (de Souza and Smith 1989; NRC 1991c) and should be considered.

SURGERY

Appropriate attention to presurgical planning, personnel training, aseptic and surgical technique, animal well-being, and animal physiologic status during all phases of a protocol will enhance the outcome of surgery (see Appendix A, "Anesthesia, Pain, and Surgery"). The individual impact of those factors will vary according to the complexity of procedures involved and the species of animal used. A team approach to a surgical project often increases the likelihood of a successful outcome by providing input from persons with different expertise (Brown and Schofield 1994; Brown and others 1993).

A continuing and thorough assessment of surgical outcomes should be performed to ensure that appropriate procedures are followed and timely corrective changes instituted. Modification of standard techniques might be desirable or even required (for instance, in rodent or field surgery), but it should not compromise the well-being of the animals. In the event of modification, assessment of outcomes should be even more intense and might have to incorporate criteria other than obvious clinical morbidity and mortality.

Presurgical planning should include input from all members of the surgical team, including the surgeon, anesthetist, veterinarian, surgical technicians, animal-care staff, and investigator. The surgical plan should identify personnel, their roles and training needs, and equipment and supplies required for the procedures planned (Cunliffe-Beamer 1993); the location and nature of the facilities in which the procedures will be conducted; and preoperative animal-health assessment and postoperative care (Brown and Schofield 1994). If a nonsterile part of an animal, such as the gastrointestinal tract, is to be surgically exposed or if a procedure is likely to cause immunosuppression, preoperative antibiotics might be appropriate (Klement and others 1987). However, the use of antibiotics should never be considered as a replacement for aseptic procedures.

It is important that persons have had appropriate training to ensure that good surgical technique is practiced, that is, asepsis, gentle tissue handling, minimal dissection of tissue, appropriate use of instruments, effective hemostasis, and correct use of suture materials and patterns (Chaffee 1974; Wingfield 1979). People performing and assisting in surgical procedures in a research setting often have a wide range of educational backgrounds and might require various levels and kinds of training before they participate in surgical procedures on animals. For example, persons trained in human surgery might need training in interspecies variations in anatomy, physiology, and the effects of anesthetic and analgesic drugs, or in postoperative requirements. Training guidelines for research surgery commensurate with a person's background are available (ASR 1989) to assist institutions in developing appropriate training programs. The PHS Policy and the AWRs place responsibility with the IACUC for determining that personnel performing surgical procedures are appropriately qualified and trained in the procedures to be performed.

In general, surgical procedures are categorized as major or minor and in the laboratory setting can be further divided into survival and nonsurvival. Major survival surgery penetrates and exposes a body cavity or produces substantial impairment of physical or physiologic functions (such as laparotomy, thoracotomy, craniotomy, joint replacement, and limb amputation). Minor survival surgery does not expose a body cavity and causes little or no physical impairment (such as wound suturing; peripheral-vessel cannulation; such routine farm-animal procedures as castration, dehorning, and repair of prolapses; and most procedures routinely done on an "outpatient" basis in veterinary clinical practice).

Minor procedures are often performed under less-stringent conditions than major procedures but still require aseptic technique and instruments and appropriate anesthesia. Although laparoscopic procedures are often performed on an "outpatient" basis, appropriate aseptic technique is necessary if a body cavity is penetrated.

In nonsurvival surgery, an animal is euthanatized before recovery from anesthesia. It might not be necessary to follow all the techniques outlined in this section if nonsurvival surgery is performed; however, at a minimum, the surgical site should be clipped, the surgeon should wear gloves, and the instruments and surrounding area should be clean (Slattum and others 1991).

Emergency situations sometimes require immediate surgical correction under less than ideal conditions. For example, if an animal maintained outdoors needs surgical attention, movement to a surgical facility might pose an unacceptable risk to the animal or be impractical. Such situations often require more-intensive aftercare and might pose a greater risk of postoperative complications. The appropriate course of action requires veterinary medical judgment.

Aseptic technique is used to reduce microbial contamination to the lowest possible practical level (Cunliffe-Beamer 1993). No procedure, piece of equipment, or germicide alone can achieve that objective (Schonholtz 1976). Aseptic technique requires the input and cooperation of everyone who enters the operating suite (Belkin 1992; McWilliams 1976). The contribution and importance of each practice varies with the procedure. Aseptic technique includes preparation of the patient, such as hair removal and disinfection of the operative site (Hofmann 1979); preparation of the surgeon, such as the provision of decontaminated surgical attire, surgical scrub, and sterile surgical gloves (Chamberlain and Houang 1984; Pereira and others 1990; Schonholtz 1976); sterilization of instruments, supplies, and implanted materials (Kagan 1992b); and the use of operative techniques to reduce the likelihood of infection (Ayliffe 1991; Kagan 1992a; Ritter and Marmion 1987; Schofield 1994; Whyte 1988).

Specific sterilization methods should be selected on the basis of physical characteristics of materials to be sterilized (Schofield 1994). Autoclaving and gas sterilization are common effective methods. Sterilization indicators should be used to identify materials that have undergone proper sterilization (Berg 1993). Liquid chemical sterilants should be used with adequate contact times, and instruments should be rinsed with sterile water or saline before use. Alcohol is neither a sterilant nor a high-level disinfectant (Rutala 1990).

In general, unless an exception is specifically justified as an essential component of the research protocol and approved by the IACUC, nonrodent aseptic surgery should be conducted only in facilities intended for that purpose. Most bacteria are carried on airborne particles or fomites, so surgical facilities should be maintained and operated in a manner that ensures cleanliness and minimizes unnecessary traffic (AORN 1982; Bartley 1993). In some circumstances, it might be necessary to use an operating room for other purposes. In such cases, it is imperative that the room be returned to an appropriate level of cleanliness before its use for major survival surgery.

Careful surgical monitoring and timely attention to problems increase the likelihood of a successful surgical outcome. Monitoring includes checking of anesthetic depth and physiologic function and assessment of clinical signs and conditions. Maintenance of normal body temperature minimizes cardiovascular and respiratory disturbances caused by anesthetic agents (Dardai and Heavner 1987) and is of particular importance.

The species of animal influences the components and intensity of the surgical program. The relative susceptibility of rodents to surgical infection has been debated; available data suggest that subclinical infections can cause adverse physiologic and behavioral responses (Beamer 1972; Bradfield and others 1992; Cunliffe-Beamer 1990; Waynforth 1980, 1987) that can affect both surgical success and research results. Some characteristics of common laboratory-rodent surgery—such as smaller incision sites, fewer personnel in the surgical team, manipulation of multiple animals at one sitting, and briefer procedures—as opposed to surgery in larger species, can make modifications in standard aseptic techniques necessary or desirable (Brown 1994; Cunliffe-Beamer 1993). Useful suggestions for dealing with some of the unique challenges of rodent surgery have been published (Cunliffe-Beamer 1983, 1993).

Generally, farm animals maintained for biomedical research should undergo surgery with procedures and in facilities compatible with the guidelines set forth in this section. However, some minor and emergency procedures that are commonly performed in clinical veterinary practice and in commercial agricultural settings may be conducted under less-stringent conditions than experimental surgical procedures in a biomedical-research setting. Even when conducted in an agricultural setting, these procedures require the use of appropriate aseptic technique, sedatives, analgesics, anesthetics, and conditions commensurate with the risk to the animal's health and well-being. But they might not require the intensive surgical settings, facilities, and procedures outlined here.

Presurgical planning should specify the requirements of postsurgical monitoring, care, and record-keeping, including the personnel who will perform these duties. The investigator and veterinarian share responsibility for ensuring that postsurgical care is appropriate. An important component of postsurgical care is observation of the animal and intervention as required during recovery from anesthesia and surgery. The intensity of monitoring necessary will vary with the species and the procedure and might be greater during the immediate anesthetic-recovery period than later in postoperative recovery. During the anesthetic-recovery period, the animal should be in a clean, dry area where it can be observed often by trained personnel. Particular attention should be given to thermoregulation, cardiovascular and respiratory function, and postoperative pain or discomfort during recovery from anesthesia. Additional care might be warranted, including administration of parenteral fluids for maintenance of water and electrolyte balance (FBR 1987), analgesics, and other drugs; care for surgical incisions; and maintenance of appropriate medical records.

After anesthetic recovery, monitoring is often less intense but should include attention to basic biologic functions of intake and elimination and behavioral signs of postoperative pain, monitoring for postsurgical infections, monitoring of the surgical incision, bandaging as appropriate, and timely removal of skin sutures, clips, or staples (UFAW 1989).

PAIN, ANALGESIA, AND ANESTHESIA

An integral component of veterinary medical care is prevention or alleviation of pain associated with procedural and surgical protocols. Pain is a complex experience that typically results from stimuli that damage tissue or have the potential to damage tissue. The ability to experience and respond to pain is widespread in the animal kingdom. A painful stimulus prompts withdrawal and evasive action. Pain is a stressor and, if not relieved, can lead to unacceptable levels of stress and distress in animals. The proper use of anesthetics and analgesics in research animals is an ethical and scientific imperative. *Recognition and Alleviation of Pain and Distress in Laboratory Animals* (NRC 1992) is a source of information about the basis and control of pain (see also Appendix A).

Fundamental to the relief of pain in animals is the ability to recognize its clinical signs in specific species (Hughes and Lang 1983; Soma 1987). Species vary in their response to pain (Breazile 1987; Morton and Griffiths 1985; Wright and others 1985), so criteria for assessing pain in various species differ. Some species-specific behavioral manifestations of pain or distress are used as indicators, for example, vocalization, depression or other behavioral changes, abnormal appearance or posture, and immobility (NRC 1992). It is therefore essential that personnel caring for and using animals be very familiar with species-specific (and individual) behavioral, physiologic, and biochemical indicators of well-being (Dresser 1988; Dubner 1987; Kitchen and others 1987). In general, unless the contrary is known or established it should be assumed that procedures that cause pain in humans also cause pain in animals (IRAC 1985).

The selection of the most appropriate analgesic or anesthetic should reflect professional judgment as to which best meets clinical and humane requirements without compromising the scientific aspects of the research protocol. Preoperative or intraoperative administration of analgesics might enhance postsurgical analgesia. The selection depends on many factors, such as the species and age of the animal, the type and degree of pain, the likely effects of particular agents on specific organ systems, the length of the operative procedure, and the safety of an agent for an animal, particularly if a physiologic deficit is induced by a surgical or other experimental procedure. Such devices as precision vaporizers and respirators increase the safety and choices of inhalation agents for use in rodents and other small species.

Some classes of drugs—such as sedatives, anxiolytics, and neuromuscular blocking agents—are not analgesic or anesthetic and thus do not relieve pain; however, they might be used in combination with appropriate analgesics and anesthetics. Neuromuscular blocking agents (e.g., pancuronium) are sometimes used to paralyze skeletal muscles during surgery in which general anesthetics have been administered (Klein 1987). When these agents are used during surgery or in any other painful procedure, many signs of anesthetic depth are eliminated because of the paralysis. However, autonomic nervous system changes (e.g., sudden changes in heart rate and blood pressure) can be indicators of pain related to an inadequate depth of anesthesia. If paralyzing agents are to be used, it is recommended that the appropriate amount of anesthetic be first defined on the basis of results of a similar procedure that used the anesthetic without a blocking agent (NRC 1992).

In addition to anesthetics, analgesics, and tranquilizers, nonpharmacologic control of pain is often effective (NRC 1992; Spinelli 1990).

Neuromuscular blocking drugs, as noted earlier, do not provide relief from pain. They are used to paralyze skeletal muscles while an animal is fully anesthetized. They might be used in properly ventilated conscious animals for specific types of nonpainful, well-controlled neurophysiologic studies. However, it is imperative that any such proposed use be carefully evaluated by the IACUC to ensure the well-being of the animal because acute stress is believed to be a consequence of paralysis in a conscious state and it is known that humans, if conscious, can experience distress when paralyzed with these drugs (NRC 1992; Van Sluyters and Oberdorfer 1991).

EUTHANASIA

Euthanasia is the act of killing animals by methods that induce rapid unconsciousness and death without pain or distress. Unless a deviation is justified for scientific or medical reasons, methods should be consistent with the *1993 Report of the AVMA Panel on Euthanasia* (AVMA 1993 or later editions). In evaluating the appropriateness of methods, some of the criteria that should be considered are ability to induce loss of consciousness and death with no or only momentary pain, distress, or anxiety; reliability; nonreversibility; time required to induce unconsciousness; species and age limitations; compatibility with research objectives; and safety of and emotional effect on personnel.

Euthanasia might be necessary at the end of a protocol or as a means to relieve pain or distress that cannot be alleviated by analgesics, sedatives, or other treatments. Protocols should include criteria for initiating euthanasia, such as degree of a physical or behavioral deficit or tumor size, that will enable a prompt decision to be made by the veterinarian and the investigator to ensure that the end point is humane and the objective of the protocol is achieved.

Euthanasia should be carried out in a manner that avoids animal distress. In some cases, vocalization and release of pheromones occur during induction of unconsciousness. For that reason, other animals should not be present when euthanasia is performed (AVMA 1993).

The selection of specific agents and methods for euthanasia will depend on the species involved and the objectives of the protocol. Generally, inhalant or noninhalant chemical agents (such as barbiturates, nonexplosive inhalant anesthetics, and CO_2) are preferable to physical methods (such as cervical dislocation, decapitation, and use of a penetrating captive bolt). However, scientific considerations might preclude the use of chemical agents for some protocols. All methods of euthanasia should be reviewed and approved by the IACUC.

It is essential that euthanasia be performed by personnel who are skilled in methods for the species in question and that it be performed in a professional and compassionate manner. Death should be confirmed by personnel who can recognize cessation of vital signs in the species being euthanatized. Euthanatizing animals is psychologically difficult for some animal-care, veterinary, and research personnel, particularly if they are involved in performing euthanasia repetitively or if they have become emotionally attached to the animals being euthanatized (Arluke 1990; NRC 1992; Rollin 1986; Wolfle 1985). When delegating euthanasia responsibilities, supervisors should be aware of this as a potential problem for some employees or students.

REFERENCES

Arluke, A. 1990. Uneasiness among laboratory technicians. Lab. Anim. 19(4):20-39.

AORN (Association of Operating Room Nurses). 1982. Recommended practices for traffic patterns in the surgical suite. Assoc. Oper. Room Nurs. J. 15(4):750-758.

ASR (Academy of Surgical Research). 1989. Guidelines for training in surgical research in animals. J. Invest. Surg. 2:263-268.

Ayliffe, G. A. J. 1991. Role of the environment of the operating suite in surgical wound infection. Rev. Inf. Dis. 13(Suppl 10):S800-804.

AVMA (American Veterinary Medical Association). 1993. Report of the AVMA panel on euthanasia. J. Am. Vet. Med. Assoc. 202(2):229-249.

Bartley, J. M. 1993. Environmental control: Operating room air quality. Today's O.R. Nurse 15(5):11-18.

Beamer, T. C. 1972. Pathological changes associated with ovarian transplantation. Pp. 104 in The 44th Annual Report of the Jackson Laboratory, Bar Harbor, Maine: Jackson Laboratory.

Belkin, N. J. 1992. Barrier materials, their influence on surgical wound infections. Assoc. Oper. Room Nurs. J. 55(6):1521-1528.

Berg, J. 1993. Sterilization. Pp. 124-129 in Textbook of Small Animal Surgery, 2nd ed., D. Slatter, ed. Philadelphia: W. B. Saunders.

Bradfield, J. F., T. R. Schachtman, R. M. McLaughlin, and E. K. Steffen. 1992. Behavioral and physiological effects of inapparent wound infection in rats. Lab. Anim. Sci. 42(6):572-578.

Breazile, J. E. 1987. Physiologic basis and consequences of distress in animals. J. Am. Vet. Med. Assoc. 191(10):1212-1215.

Brown, M. J. 1994. Aseptic surgery for rodents. Pp. 67-72 in Rodents and Rabbits: Current Research Issues, S. M. Niemi, J. S. Venable, and H. N. Guttman, eds. Bethesda, Md.: Scientists Center for Animal Welfare.

Brown, M. J., and J. C. Schofield. 1994. Perioperative care. Pp. 79-88 in Essentials for Animal Research: A Primer for Research Personnel. B. T. Bennett, M. J. Brown, and J. C. Schofield, eds. Washington, D. C.: National Agricultural Library.

Brown, M. J., P. T. Pearson, and F. N. Tomson. 1993. Guidelines for animal surgery in research and teaching. Am. J. Vet. Res. 54(9):1544-1559.

Butler, T. M., B. G. Brown, R. C. Dysko, E. W. Ford, D. E. Hoskins, H. J. Klein, J. L. Levin, K. A. Murray, D. P. Rosenberg, J. L. Southers, and R. B. Swenson. 1995. Medical management. Pp. 255-334 in Nonhuman Primates in Biomedical Research: Biology and Management, B. T. Bennett, C. R. Abee, and R. Hendrickson, eds. San Diego, Calif.: Academic Press.

CDC (Centers for Disease Control and Prevention). 1991. Update: Nonhuman primate importation. MMWR, October 9, 1991.

CDC (Centers for Disease Control and Prevention). 1993. Tuberculosis in imported nonhuman primates-United States, June 1990-May 1993. MMWR, July 30, 1993. Vol. 42, no. 29.

CFR (Code of Federal Regulations) Title 42. PHS, HHS, Subchapter F (Importations), Section 71.53 (Nonhuman primates).

Chaffee, V. W. 1974. Surgery of laboratory animals. Pp. 233-247 in Handbook of Laboratory Animal Science, Vol. 1, E. C. Melby, Jr. and N. H. Altman, eds. Cleveland, Ohio: CRC Press.

Chamberlain, G. V., and E. Houang. 1984. Trial of the use of masks in gynecological operating theatre. Ann. R. Coll. Surg. 66(6):432-433.

Cunliffe-Beamer, T. L. 1983. Biomethodology and surgical techniques. Pp. 419-420 in The Mouse in Biomedical Research, Vol. III, Normative Biology, Immunology and Husbandry. H. L. Foster, J. D. Small and J. G. Fox, eds. New York: Academic Press.

Cunliffe-Beamer, T. L. 1990. Surgical Techniques. Pp. 80-85 in Guidelines for the Well-Being of Rodents in Research, H. N. Guttman, ed. Bethesda, Md.: Scientists Center for Animal Welfare.

Cunliffe-Beamer, T. L. 1993. Applying principles of aseptic surgery to rodents. AWIC Newsl. 4(2):3-6.

Dardai, E., and J. E. Heavner. 1987. Respiratory and cardiovascular effects of halothane, isoflurane and enflurane delivered via a Jackson-Rees breathing system in temperature controlled and uncontrolled rats. Meth. Find. Exp. Clin. Pharmacol. 9(11):717-720.

de Souza, M., and A. L. Smith. 1989. Comparison of isolation in cell culture with conventional and modified mouse antibody production tests for detection of murine viruses. J. Clin. Microbiol. 27:185-187.

DOI (Department of the Interior). Endangered and threatened wildlife and plants (50 CFR 17.11), U.S. Fish and Wildlife Service.

Dresser, R. 1988. Assessing harm and justification in animal research: Federal policy opens the laboratory door. Rutgers Law Rev. 450(3):723-795.

Drozdowicz, C. K., T. A. Bowman, M. L. Webb, and C. M. Lang. 1990. Effect of in-house transport on murine plasma corticosterone concentration and blood lymphocyte populations. Am. J. Vet. Res. 51:1841-1846.

Dubner, R. 1987. Research on pain mechanisms in animals. J. Am. Vet. Med. Assoc. 191(10):1273-1276.

FBR (Foundation for Biomedical Research). 1987. Surgery: Protecting your animals and your study. Pp. 19-27 in The Biomedical Investigator's Handbook for Researchers Using Animal Models. Washington, D. C.: Foundation for Biomedical Research.

FR (Federal Register) 1990. CDC, HHS. Requirement for a special permit to import cynomolgus, African green, or rhesus monkeys into the United States, Vol. 55, no. 77, April 20, 1990.

Hirsch, V. M., P. M. Zack, A. P. Vogel, and P. R. Johnson. 1991. Simian immunodeficiency virus infection of macaques: End-stage disease is characterized by wide-spread distribution of proviral DNA in tissues. J. Infect. Dis. 163:976-988.

Hofmann, L. S. 1979. Preoperative and operative patient management. Pp. 14-22 in Small Animal Surgery, An Atlas of Operative Technique, W. E. Wingfield and C. A. Rawlings, eds. Philadelphia: W. B. Saunders.

Holmes, A. W., R. G. Caldwell, R. E. Dedmon, and F. Deinhardt. 1964. Isolation and characterization of a new herpes virus. J. Immunol. 92:602-610.

Hughes, H. C., and C. M. Lang. 1983. Control of pain in dogs and cats. Pp. 207-216 in Animal Pain: Perception and Alleviation, R. L. Kitchell and H. H. Erickson, eds. Bethesda, Md.: American Physiological Society.

Hunt, R. D., and L. V. Melendez. 1966. Spontaneous herpes-T infection in the owl monkey (Aotus trivirgatus). Pathol. Vet. 3:1-26.

IATA (International Air Transport Association). 1995. IATA Live Animal Regulations, 22nd edition. Montreal, Quebec: International Air Transport Association.

IRAC (Interagency Research Animal Committee). 1985. U.S. Government Principles for Utilization and Care of Vertebrate Animals Used in Testing, Research, and Training. Federal Register, May 20, 1985. Washington, D.C.: Office of Science and Technology Policy.

Jelinek, V. 1971. The influence of the condition of the laboratory animals employed on the experimental results. Pp. 110-120 in Defining the Laboratory Animal.. Washington, D.C.: National Academy of Sciences.

Kagan, K. G. 1992a. Aseptic technique. Vet. Tech. 13(3):205-210.

Kagan, K. G. 1992b. Care and sterilization of surgical equipment. Vet. Tech. 13(1):65-70.

Kitchen, H., A. Aronson, J. L. Bittle, C. W. McPherson, D. B. Morton, S. P. Pakes, B. Rollin, A. N. Rowan, J. A. Sechzer, J. E. Vanderlip, J. A. Will, A. S. Clark, and J. S. Gloyd. 1987. Panel report of the colloquium on recognition and alleviation of animal pain and distress. J. Am. Vet. Med. Assoc. 191(10):1186-1191.

Klein, L. 1987. Neuromuscular blocking agents. Pp. 134-153 in Principles and Practice of Veterinary Anesthesia, C. E. Short, ed. Baltimore, Md.: Williams & Wilkins.

Klement, P., P. J. del Nido, L. Mickleborough, C. MacKay, G. Klement, and G. J. Wilson. 1987. Techniques and postoperative management for successful cardiopulmonary bypass and open-heart surgery in dogs. J. Am. Vet. Med. Assoc. 190(7):869-874.

Landi, M. S., J. W. Kreider, C. M. Lang, and L. P. Bullock. 1982. Effects of shipping on the immune function in mice. Am. J. Vet. Res. 43:1654-1657.

Manning, P. J., J. E. Wagener, and J. E. Harkness. 1984. Biology and diseases of guinea pigs. In Laboratory Animal Medicine. J. G. Fox, B. J. Cohen, and F. M. Loew, eds. San Diego: Academic Press.

McWilliams, R. M. 1976. Divided responsibilities for operating room asepsis: The dilemma of technology. Med. Instrum. 10(6):300-301.

Melnick, F. L., M. Midulla, I. Wimberly, J. G. Barrera-Oro, and B. M. Levy. 1964. A new member of the herpes virus group isolated from South American marmosets. J. Immunol. 92:596-601.

Morton, D. B., and P. H. M. Griffiths. 1985. Guidelines on the recognition of pain, distress and discomfort in experimental animals and an hypothesis for assessment. Vet. Rec. 116:431-436.

Murphey-Corb, M., L. N. Martin, S. R. S. Rangan, G. B. Baskin, B. J. Gormus, R. H. Wolf, W. A. Andes, M. West, and R. C. Montelaro. 1986. Isolation of an HTLV-III-related retrovirus from macaques with simian AIDS and its possible origin in asymptomatic managabeys. Nature 321:435-437.

Nicklas, W., V. Kraft, and B. Meyer. 1993. Contamination of transplantable tumors, cell lines, and monoclonal antibodies with rodent viruses. Lab. Anim. Sci. 43:296-299.

NRC (National Research Council). 1991a. Barrier programs. Pp. 17-20 in Infectious Diseases of

Mice and Rats. A report of the Institute of Laboratory Animal Resources Committee on Infectious Diseases of Mice and Rats. Washington, D.C.: National Academy Press.

NRC (National Research Council). 1991b. Individual disease agents and their effects on research. Pp. 31-258 in Infectious Diseases of Mice and Rats. A report of the Institute of Laboratory Animal Resources Committee on Infectious Diseases of Mice and Rats. Washington, D.C.: National Academy Press.

NRC (National Research Council). 1991c. Health Surveillance Programs. Pp. 21-27 in Infectious Diseases of Mice and Rats. A report of the Institute of Laboratory Animal Resources Committee on Infectious Diseases of Mice and Rats. Washington, D.C.: National Academy Press.

NRC (National Research Council). 1992. Recognition and Alleviation of Pain and Distress in Laboratory Animals. A report of the Institute of Laboratory Animal Resources Committee on Pain and Distress in Laboratory Animals. Washington, D.C.: National Academy Press.

Palmer, A. E., A. M. Allen, N. M. Tauraso, and A. Skelokov. 1968. Simian hemorrhagic fever. I. Clinical and epizootiologic aspects of an outbreak among quarantined monkeys. Am. J. Trop. Med. Hyg. 17:404-412.

Pereira, L. J., G. M. Lee, and K. J. Wade. 1990. The effect of surgical handwashing routines on the microbial counts of operating room nurses. Am. J. Inf. Control. 18(6):354-364.

PHS (Public Health Service). 1996. Public Health Service Policy on Humane Care and Use of Laboratory Animals. Washington, D.C.: U.S. Department of Health and Human Services, 28 pp. [PL 99-158, Health Research Extension Act, 1985]

Prasad, S., B. R. Gatmaitan, and R. C. O'Connell. 1978. Effect of a conditioning method on general safety test in guinea pigs. Lab. Anim. Sci. 28(5):591-593.

Ritter, M. A., and P. Marmion. 1987. The exogenous sources and controls of microorganisms in the operating room. Orthopaedic Nursing 7(4):23-28.

Rollin, B. 1986. Euthanasia and moral stress. In Loss, Grief and Care, R. DeBellis, ed. Binghamton, N.Y.: Haworth Press.

Rutala, W. A. 1990. APIC guideline for selection and use of disinfectants. Am. J. Inf. Control 18(2):99-117.

Sanhouri A. A., R. S. Jones, and H. Dobson. 1989. The effects of different types of transportation on plasma cortisol and testosterone concentrations in male goats. Br. Vet. J. 145:446-450.

Schofield, J. C. 1994. Principles of aseptic technique. Pp. 59-77 in Essentials for Animal Research: A Primer for Research Personnel, B. T. Bennett, M. J. Brown, and J. C. Schofield, eds. Washington, D.C.: National Agricultural Library.

Schonholtz, G. J. 1976. Maintenance of aseptic barriers in the conventional operating room. J. Bone and Joint Surg. 58-A(4):439-445.

Slattum, M. M., L. Maggio-Price, R. F. DiGiacomo, and R. G. Russell. 1991. Infusion-related sepsis in dogs undergoing acute cardiopulmonary surgery. Lab. Anim. Sci. 41(2):146-150.

Soma, L. R. 1987. Assessment of animal pain in experimental animals. Lab. Anim. Sci. 37:71-74.

Spinelli, J. 1990. Preventive suffering in laboratory animals. Pp. 231-242 in The Experimental Animal in Biomedical Research. Vol. I: A Survey of Scientific and Ethical Issues for Investigators. B. Rollin and M. Kesel, eds. Boca Raton, Fla.: CRC Press.

Tuli, J. S., J. A. Smith, and D. B. Morton. 1995. Stress measurements in mice after transportation. Lab. Anim. 29:132-138.

UFAW (Universities Federation for Animal Welfare). 1989. Surgical procedures. Pp. 3-15 in Guidelines on the Care of Laboratory Animals and Their Use for Scientific Purposes III. London: Universities Federation for Animal Welfare.

Van Sluyters, R. C., and M. D. Oberdorfer, eds. 1991. Preparation and Maintenance of Higher Mammals During Neuroscience Experiments. Report of National Institute of Health Workshop. NIH No. 91-3207. Bethesda, Md.: National Institutes of Health.

Wallace, M. E. 1976. Effect of stress due to deprivation and transport in different genotypes of house mouse. Lab. Anim. (London) 10(3):335-347.

Waynforth, H. B. 1980. Experimental and Surgical Technique in the Rat. London: Academic Press. 104 pp.

Waynforth, H. B. 1987. Standards of surgery for experimental animals. Pp. 311-312 in Laboratory Animals: An Introduction for New Experimenters, A. A. Tuffery, ed. Chichester: Wiley-Interscience.

Whyte, W. 1988. The role of clothing and drapes in the operating room. J. Hosp. Inf. 11(Suppl C):2-17.

Wingfield, W. E. 1979. Surgical Principles. Pp. 1-3 in Small Animal Surgery, An Atlas of Operative Techniques, W. E. Wingfield and C. A. Rawlings, eds. Philadelphia: W. B. Saunders.

Wolfle, T. L. 1985. Laboratory animal technicians: Their role in stress reduction and human-companion animal bonding. Vet. Clin. N. Am. Small Anim. Pract. 15(2):449-454.

Wright, E. M., K. L. Marcella, and J. F. Woodson. 1985. Animal pain: Evaluation and control. Lab Anim. 14(4):20-36.

4

Physical Plant

A well-planned, well-designed, well-constructed, and properly maintained facility is an important element of good animal care and use, and it facilitates efficient, economical, and safe operation (see Appendix A, "Design and Construction of Animal Facilities"). The design and size of an animal facility depend on the scope of institutional research activities, the animals to be housed, the physical relationship to the rest of the institution, and the geographic location. Effective planning and design should include input from personnel experienced with animal-facility design and operation and from representative users of the proposed facility. Computational fluid dynamics (CFD) modeling of new facilities and caging might be beneficial (Reynolds and Hughes 1994). An animal facility should be designed and constructed in accord with all applicable state and local building codes. Modular units (such as custom-designed trailers or prefabricated structures) should comply with construction guidelines described in this chapter.

Good animal management and human comfort and health protection require separation of animal facilities from personnel areas, such as offices and conference rooms. Separation can be accomplished by having the animal quarters in a separate building, wing, floor, or room. Careful planning should make it possible to place animal-housing areas next to or near research laboratories but separated from them by barriers, such as entry locks, corridors, or floors. Animals should be housed in facilities dedicated to or assigned for that purpose and should not be housed in laboratories merely for convenience. If animals must be maintained in a laboratory area to satisfy a protocol, the area should be appropriate to house and care for the animals; if needed, measures should be taken to minimize occupational hazards related to exposure to animals.

Building materials should be selected to facilitate efficient and hygienic operation of animal facilities. Durable, moisture-proof, fire-resistant, seamless materials are most desirable for interior surfaces. Surfaces should be highly resistant to the effects of cleaning agents, scrubbing, high-pressure sprays, and impact. Paints and glazes should be nontoxic if used on surfaces with which animals will have direct contact. In the construction of outdoor facilities, consideration should be given to surfaces that withstand the elements and can be easily maintained.

FUNCTIONAL AREAS

Professional judgment should be exercised in the development of a practical, functional, and efficient physical plant for animal care and use. The size, nature, and intensity of an institutional animal program will determine the specific facility and support functions needed. In facilities that are small, maintain few animals, or maintain animals under special conditions—such as facilities used exclusively for housing gnotobiotic or specific-pathogen-free (SPF) colonies or animals in runs, pens, or outdoor housing—some functional areas listed below might be unnecessary or might be included in a multipurpose area.

Space is required for

- Animal housing, care, and sanitation.
- Receipt, quarantine, and separation of animals.
- Separation of species or isolation of individual projects when necessary.
- Storage.

Most multipurpose animal facilities also include the following:

- Specialized laboratories or space contiguous with or near animal-housing areas for such activities as surgery, intensive care, necropsy, radiography, preparation of special diets, experimental procedures, clinical treatment, and diagnostic laboratory procedures.
- Containment facilities or equipment, if hazardous biologic, physical, or chemical agents are to be used.
- Receiving and storage areas for food, bedding, pharmaceuticals, biologics, and supplies.
- Space for washing and sterilizing equipment and supplies and, depending on the volume of work, machines for washing cages, bottles, glassware, racks, and waste cans; a utility sink; an autoclave for equipment, food, and bedding; and separate areas for holding soiled and clean equipment.
- Space for storing wastes before incineration or removal.

- Space for cold storage or disposal of carcasses.
- Space for administrative and supervisory personnel, including space for training and education of staff.
- Showers, sinks, lockers, toilets, and break areas for personnel.
- Security features, such as card-key systems, electronic surveillance, and alarms.

CONSTRUCTION GUIDELINES

Corridors

Corridors should be wide enough to facilitate the movement of personnel and equipment. Corridors 6-8 ft wide can accommodate the needs of most facilities. Floor-wall junctions should be designed to facilitate cleaning. In corridors leading to dog and swine housing facilities, cage-washing facilities, and other high-noise areas, double-door entry or other noise traps should be considered. Wherever possible, water lines, drainpipes, electric-service connections, and other utilities should be accessible through access panels or chases in corridors outside the animal rooms. Fire alarms, fire extinguishers, and telephones should be recessed or installed high enough to prevent damage from the movement of large equipment.

Animal-Room Doors

For safety, doors should open into animal rooms; however, if it is necessary that they open toward a corridor, there should be recessed vestibules. Doors with viewing windows might be preferable for safety and other reasons. However, the ability to cover viewing windows might be considered in situations where exposure to light or hallway activities would be undesirable. Doors should be large enough (approximately 42×84 in) to allow the easy passage of racks and equipment. Doors should fit tightly within their frames, and both doors and frames should be appropriately sealed to prevent vermin entry or harborage. Doors should be constructed of and, where appropriate, coated with materials that resist corrosion. Self-closing doors equipped with recessed or shielded handles, threshold sweeps, and kickplates are usually preferred. Where room-level security is necessary or it is desirable to limit access (as in the case of the use of hazardous agents), room doors should be equipped with locks. Doors should be designed to be opened from the inside without a key.

Exterior Windows

Windows are acceptable in some animal rooms and can constitute a type of environmental enrichment for some species, especially nonhuman primates, dogs, some agricultural animals, and other large mammals. The effects of windows on temperature, photoperiod control, and security should be considered in design decisions. Where temperature cannot be regulated properly because of heat loss or gain through the windows or where photoperiod is an important consideration (as in breeding colonies of rodents), exterior windows usually are inappropriate.

Floors

Floors should be moisture-resistant, nonabsorbent, impact-resistant, and relatively smooth, although textured surfaces might be required in some high-moisture areas and for some species (such as farm animals). Floors should be resistant to the action of urine and other biologic materials and to the adverse effects of hot water and cleaning agents. They should be capable of supporting racks, equipment, and stored items without becoming gouged, cracked, or pitted. Depending on their use, floors should be monolithic or have a minimal number of joints. Some materials that have proved satisfactory are epoxy aggregates, hard-surface sealed concrete, and special hardened rubber-base aggregates. Correct installation is essential to ensure long-term stability of the surface. If sills are installed at the entrance to a room, they should be designed to allow for convenient passage of equipment.

Drainage

Where floor drains are used, the floors should be sloped and drain traps kept filled with liquid. To minimize humidity, drainage should allow rapid removal of water and drying of surfaces (Gorton and Besch 1974). Drainpipes should be at least 4 in (10.2 cm) in diameter. In some areas, such as dog kennels and farm-animal facilities, larger drain pipes are recommended. A rim-flush drain or heavy-duty disposal unit set in the floor might be useful for the disposal of solid waste. When drains are not in use for long periods, they should be capped and sealed to prevent backflow of sewer gases and other contaminants; lockable drain covers might be advisable for this purpose in some circumstances.

Floor drains are not essential in all animal rooms, particularly those housing rodents. Floors in such rooms can be sanitized satisfactorily by wet vacuuming or mopping with appropriate cleaning compounds or disinfectants.

Walls

Walls should be smooth, moisture-resistant, nonabsorbent, and resistant to damage from impact. They should be free of cracks, of unsealed utility penetrations, and of imperfect junctions with doors, ceilings, floors, and corners. Surface materials should be capable of withstanding cleaning with detergents and disinfectants and the impact of water under high pressure. The use of curbs, guardrails or bumpers, and corner guards should be considered to protect walls and corners from damage.

Ceilings

Ceilings should be smooth, moisture-resistant, and free of imperfect junctions. Surface materials should be capable of withstanding cleaning with detergents and disinfectants. Ceilings of plaster or fire-proof plasterboard should be sealed and finished with a washable paint. Ceilings formed by the concrete floor above are satisfactory if they are smoothed and sealed or are painted. Generally, suspended ceilings are undesirable unless they are fabricated of impervious materials and free of imperfect junctions. Exposed plumbing, ductwork, and light fixtures are undesirable unless the surfaces can be readily cleaned.

Heating, Ventilation, and Air-Conditioning (HVAC)

Temperature and humidity control minimizes variations due either to changing climatic conditions or to differences in the number and kind of animals in a room. Air-conditioning is an effective means of regulating temperature and humidity. HVAC systems should be designed for reliability, ease of maintenance, and energy conservation. They should be able to meet requirements for animals as discussed in Chapter 2. A system should be capable of adjustments in dry-bulb temperatures of ±1°C (±2°F). The relative humidity should generally be maintained within a range of 30-70% throughout the year. Temperature is best regulated by having thermostatic control for each room. Use of a zonal control for multiple rooms can result in temperature variations between the "master-control" animal room and the other rooms in the zone because of differences in animal densities within the rooms and heat gain or loss in ventilation ducts and other surfaces within the zone.

Regular monitoring of the HVAC system is important and is best done at the individual-room level. Previously specified temperature and humidity ranges can be modified to meet special animal needs in circumstances in which all or most of the animal facility is designed exclusively for acclimated species with similar requirements (for example, when animals are held in a sheltered or outdoor facility).

Brief and infrequent, moderate fluctuations in temperature and relative humidity outside suggested ranges are well tolerated by most species commonly used in research. Most HVAC systems are designed for average high and low temperatures and humidities experienced in a geographic area within ±5% variation (ASHRAE 1993). When extremes in external ambient conditions that are beyond design specifications occur, provisions should be in place to minimize the magnitude and duration of fluctuations in temperature and relative humidity outside the recommended ranges. Such measures can include partial redundancy, partial recycling of air, altered ventilation rates, or the use of auxiliary equipment. In the event of a partial HVAC system failure, systems should be designed to supply facility needs at a reduced level. It is essential that life-threatening heat accumulation or loss be prevented during mechanical failure. Totally redundant systems are seldom practical or necessary except under special circumstances (as in some biohazard areas). Temporary needs for ventilation of sheltered or outdoor facilities can usually be met with auxiliary equipment.

In some instances, high-efficiency particulate air (HEPA) filters are recommended for air supplied to animal-holding, procedural, and surgical facilities. Also, consideration should be given to the regulation of air-pressure differentials in surgical, procedural, housing, and service areas. For example, areas for quarantine, housing, and use of animals exposed to hazardous materials and for housing of nonhuman primates should be kept under relative negative pressure, whereas areas for surgery, for clean-equipment storage, and for housing of pathogen-free animals should be kept under relative positive pressure with clean air. Maintaining air-pressure differentials is not the principal or only method by which cross contamination is controlled and should not be relied on for containment. Few air-handling systems have the necessary controls or capacity to maintain pressure differentials across doors or similar structures when they are opened for even brief periods. Containment requires the use of biologic-safety cabinets and exhausted airlocks or other means, some of which are described in Chapter 1.

If recirculated air is used, its quality and quantity should be in accord with recommendations in Chapter 2. The type and efficiency of air treatment should be matched to the quantity and types of contaminants and to the risks that they pose.

Power and Lighting

The electric system should be safe and provide appropriate lighting, a sufficient number of power outlets, and suitable amperage for specialized equipment. In the event of power failure, an alternative or emergency power supply should be available to maintain critical services (for example, the HVAC system) or support functions (for example, freezers, ventilated racks, and isolators) in animal rooms, operating suites, and other essential areas.

Light fixtures, timers, switches, and outlets should be properly sealed to prevent vermin from living there. Recessed energy-efficient fluorescent lights are most commonly used in animal facilities. A time-controlled lighting system should be used to ensure a uniform diurnal lighting cycle. Timer performance and timer-overriding systems should be checked regularly to ensure proper cycling. Light bulbs or fixtures should be equipped with protective covers to ensure the safety of the animals and personnel. Moisture-resistant switches and outlets and ground-fault interrupters should be used in areas with high water use, such as cage-washing areas and aquarium-maintenance areas.

Storage Areas

Adequate space should be provided for storage of equipment, supplies, food, bedding, and refuse. Corridors used for passage of personnel or equipment are not appropriate storage areas. Storage space can be minimized when delivery is reliable and frequent. Bedding and food should be stored in a separate area in which materials that pose a risk of contamination from toxic or hazardous substances are not stored. Refuse-storage areas should be separated from other storage areas (see Chapter 2). Refrigerated storage, separated from other cold storage, is essential for storage of dead animals and animal-tissue waste; this storage area should be kept below 7°C (44.6°F) to reduce putrefaction of wastes and animal carcasses.

Noise Control

Noise control is an important consideration in an animal facility (see Chapter 2). Noise-producing support functions, such as cage-washing, are commonly separated from housing and experimental functions. Masonry walls are more effective than metal or plaster walls in containing noise because their density reduces sound transmission. Generally, acoustic materials applied directly to the ceiling or as part of a suspended ceiling of an animal room present problems for sanitation and vermin control and are not recommended. However, sanitizable sound-attenuating materials bonded to walls or ceilings might be appropriate for noise control in some situations. Experience has shown that well-constructed corridor doors, sound-attenuating doors, or double-door entry can help to control the transmission of sound along corridors.

Attention should be paid to attenuating noise generated by equipment. Fire and environmental-monitoring alarm systems and public-address systems should be selected and located to minimize potential animal exposure. The much-higher frequencies that are capable of being discriminated by some species make it important to consider the location of equipment capable of generating sound at ultrasonic frequencies.

Facilities for Sanitizing Materials

A dedicated, central area for sanitizing cages and ancillary equipment should be provided. Mechanical cage-washing equipment is generally needed and should be selected to match the types of caging and equipment used. Consideration should be given to such factors as

• Location with respect to animal rooms and waste-disposal and storage areas.

- Ease of access, including doors of sufficient width to facilitate movement of equipment.
- Sufficient space for staging and maneuvering of equipment.
- Provision for safe bedding disposal and prewashing activities.
- Traffic flow that separates animals and equipment moving between clean and soiled areas.
- Insulation of walls and ceilings where necessary.
- Sound attenuation.
- Utilities, such as hot and cold water, steam, floor drains, and electric power.
- Ventilation, including installation of vents and provision for dissipation of steam and fumes from sanitizing processes.

FACILITIES FOR ASEPTIC SURGERY

The design of a surgical facility should accommodate the species to be operated on and the complexity of the procedures to be performed (Hessler 1991; see also Appendix A, "Design and Construction of Animal Facilities"). For most rodent surgery, a facility may be small and simple, such as a dedicated space in a laboratory appropriately managed to minimize contamination from other activities in the room during surgery. The facility often becomes larger and more complex as the number of animals, the size of animals, or the complexity of procedures increases, for instance, large-volume rodent procedures, the need for special restraint devices, hydraulic operating tables, and floor drains for farm-animal surgery, and procedures that require large surgical teams and support equipment and thus large space. The relationship of surgical facilities to diagnostic laboratories, radiology facilities, animal housing, staff offices, and so on should be considered in the overall context of the complexity of the surgical program. Surgical facilities should be sufficiently separate from other areas to minimize unnecessary traffic and decrease the potential for contamination (Humphreys 1993). Centralized facilities provide important advantages in cost savings in equipment, conservation of space and personnel resources, reduced transit of animals, and enhanced professional oversight of facilities and procedures.

For most surgical programs, functional components of aseptic surgery include surgical support, animal preparation, surgeon's scrub, operating room, and postoperative recovery. The areas that support those functions should be designed to minimize traffic flow and separate the related, nonsurgical activities from the surgical procedure in the operating room. The separation is best achieved by physical barriers (AORN 1982) but might also be achieved by distance between areas or by the timing of appropriate cleaning and disinfection between activities. The number of personnel and their level of activity have been shown to be directly related to the level of bacterial contamination and the incidence of postoperative wound infection (Fitzgerald 1979). Traffic in the operating room itself can be reduced by the installation of an observation window, a communication system (such as an intercom system), and judicious location of doors.

Control of contamination and ease of cleaning should be key considerations in the design of a surgical facility. The interior surfaces should be constructed of materials that are monolithic and impervious to moisture. Ventilation systems supplying filtered air at positive pressure can reduce the risk of postoperative infection (Ayscue 1986; Bartley 1993; Bourdillon 1946; Schonholtz 1976). Careful location of air supply and exhaust ducts and appropriate room-ventilation rates are also recommended to minimize contamination (Ayliffe 1991; Bartley 1993; Holton and Ridgway 1993; Humphreys 1993). To facilitate cleaning, the operating rooms should have as little fixed equipment as possible (Schonholtz 1976; UFAW 1989). Other features of the operating room to consider include surgical lights to provide adequate illumination (Ayscue 1986), sufficient electric outlets for support equipment, and gas-scavenging capability.

The surgical-support area should be designed for washing and sterilizing instruments and for storing instruments and supplies. Autoclaves are commonly placed in this area. It is often desirable to have a large sink in the animal-preparation area to facilitate cleaning of the animal and the operative site. A dressing area should be provided for personnel to change into surgical attire; a multipurpose locker room can serve this function. There should be a scrub area for surgeons, equipped with foot, knee, or electric-eye surgical sinks (Knecht and others 1981). To minimize the potential for contamination of the surgical site by aerosols generated during scrubbing, the scrub area is usually outside the operating room.

A postoperative-recovery area should provide the physical environment to support the needs of the animal during the period of anesthetic and immediate postsurgical recovery and should be so placed as to allow adequate observation of the animal during this period. The electric and mechanical requirements of monitoring and support equipment should be considered. The type of caging and support equipment will depend on the species and types of procedures but should be designed to be easily cleaned and to support physiologic functions, such as thermoregulation and respiration. Depending on the circumstances, a postoperative recovery area for farm animals might be modified or nonexistent in some field situations, but precautions should be taken to minimize risk of injury to recovering animals.

REFERENCES

AORN (Association of Operating Room Nurses). 1982. Recommended practices for traffic patterns in the surgical suite. Assoc. Oper. Room Nurs. J. 15(4):750-758.

ASHRAE (American Society of Heating, Refrigeration, and Air Conditioning Engineers, Inc.). 1993. Chapter 24: Weather Data. In 1993 ASHRAE Handbook: Fundamentals, I-P edition Atlanta: ASHRAE.

Ayliffe, G. A. J. 1991. Role of the environment of the operating suite in surgical wound infection. Rev. of Infec. Dis. 13(Suppl 10):S800-S804.

Ayscue, D. 1986. Operating room design: Accommodating lasers. Assoc. Oper. Room Nurs. J. 41:1278-1285.

Bartley, J. M. 1993. Environmental control: Operating room air quality. Today's O.R. Nurse 15(5):11-18.

Bourdillon, R. B. 1946. Air hygiene in dressing-rooms for burns or major wounds. The Lancet :601-605.

Fitzgerald, R. H. 1979. Microbiologic environment of the conventional operating room. Arch. Surg. 114:772-775.

Gorton, R. L., and E. L. Besch. 1974. Air temperature and humidity response to cleaning water loads in laboratory animal storage facilities. ASHRAE Trans. 80:37-52.

Hessler, J. R. 1991. Facilities to support research. Pp. 34-55 in Handbook of Facility Planning. Vol. 2: Laboratory Animal Facilities, T. Ruys, ed. New York: Van Nostrand. 422 pp.

Holton, J., and G. L. Ridgway. 1993. Commissioning operating theatres. J. Hosp. Infec. 23:153-160.

Humphreys, H. 1993. Infection control and the design of a new operating theatre suite. J. Hosp. Infec. 23:61-70.

Knecht, C. D., A. R. Allen, D. J. Williams, and J. H. Johnson. 1981. Fundamental Techniques in Veterinary Surgery, 2nd ed. Philadelphia: W. B. Saunders.

Reynolds, S. D., and H. Hughes. 1994. Design and optimization of airflow patterns. Lab Anim. 23(9):46-49.

Schonholtz, G. J. 1976. Maintenance of aseptic barriers in the conventional operating room. J. Bone and Joint Surg. 58-A(4):439-445.

UFAW (Universities Federation for Animal Welfare). 1989. Guidelines on the Care of Laboratory Animals and Their Use for Scientific Purposes: III Surgical Procedures. Herts, UK: UFAW.

APPENDIX
A

Selected Bibliography

ADMINISTRATION AND MANAGEMENT

Animal Care and Use Committees Bibliography. T. Allen and K. Clingerman. 1992. Beltsville, Md.: U.S. Department of Agriculture, National Agricultural Library (Publication #SRB92-16). 38 pp.

Animal Care and Use: Policy Issues in the 1990's. National Institutes of Health/Office for the Protection from Research Risks (NIH/OPRR). 1989. Proceedings of NIH/OPRR Conference, Bethesda, Md.

Cost Analysis and Rate Setting Manual for Animal Resource Facilities. Animal Resources Program (ARP), Division of Research Resources (DRR), National Institutes of Health (NIH). 1979 revised. NIH Pub. No. 80-2006. Washington, D.C.: U.S. Department of Health, Education and Welfare. 115 pp. (Available from ARP, DRR, NIH, Building 31, Room 5B59, Bethesda, MD 20205).

Effective Animal Care and Use Committees. F. B. Orlans, R. C. Simmonds, and W. J. Dodds, eds. 1987. In Laboratory Animal Science, Special Issue, January 1987. Published in collaboration with the Scientists Center for Animal Welfare.

Essentials for Animal Research: A Primer for Research Personnel. B. T. Bennett, M. J. Brown, and J. C. Schofield. 1994. Beltsville, Md.: National Agricultural Library. 126 pp.

Guide to the Care and Use of Experimental Animals, Volume 1, 2nd ed. E. D. Olfert, B. M. Cross, and A. A. McWilliam, eds. 1993. Ontario, Canada: Canadian Council on Animal Care. 211 pp.

Institutional Animal Care and Use Committee Guidebook. NIH/OPRR. 1992. NIH. Pub. 92-3415. (IACUC duties, special considerations, federal regulations, references and resources.)

Laboratory Animal Medical Subject Headings, A Report. NRC (National Research Council). 1972. A report of the ILAR (Institute of Laboratory Animal Resources) Committee on Laboratory Animal Literature. Washington, D.C.: National Academy of Sciences. 212 pp.

Reference Materials for Members of Animal Care and Use Committees. D. J. Berry. 1991. Beltsville, Md.: U.S. Department of Agriculture, National Agricultural Library (AWIC series #10). 42 pp.

ALTERNATIVES

Alternative Methods for Toxicity Testing: Regulatory Policy Issues. EPA-230/12-85-029. NTIS PB8-6-113404/AS. Office of Policy, Planning and Evaluation, U.S. Environmental Protection Agency. Washington, DC 20460.

Alternatives to Animal Use in Research, Testing, and Education. Office of Technology Assessment (OTA-BA-273). U.S. Gov. Printing Office. Washington, DC 20402.

Alternatives to Current Uses of Animals in Research, Safety Testing, and Education. M. L. Stephens. 1986. Washington, D.C.: Humane Society of the United States. 86 pp.

Alternatives to Pain in Experiments on Animals. D. Pratt. 1980. Argus Archives. 283 pp.

Animals and Alternatives in Testing: History, Science, and Ethics. J. Zurlo, D. Rudacile, and A. M. Goldberg. 1994. New York: Mary Ann Liebert Publishers. 86 pp.

The Principles of Humane Experimental Techniques. W. M. S. Russell and R. L. Burch. 1959. London: Methuen & Co. 238 pp. (Reprinted as a Special Edition in 1992 by the Universities Federation for Animal Welfare.)

AMPHIBIANS, REPTILES, AND FISHES

Artificial Seawaters: Formulas and Methods. J. P. Bidwell and S. Spotte. 1985. Boston: Jones and Bartlett.

The Care and Use of Amphibians, Reptiles, and Fish in Research. D. O Schaeffer, K. M. Kleinow, and L. Krulisch, eds. 1992. Proceedings from a SCAW/LSU-SVM-sponsored conference, April 8-9, 1991, New Orleans, La. Greenbelt, Md.: Scientists Center for Animal Welfare.

Disease Diagnosis and Control in North American Marine Aquaculture. 2nd rev. ed. C. J. Sindermann and D. V. Lichtner. 1988. New York: Elsevier. 426 pp.

Diseases of Fishes, Book 2A, Bacterial Diseases of Fishes. G. L. Bullock, D. A. Conroy, and S. F. Snieszko. 1971. Neptune, N.J.: T. F. H. Publications. 151 pp.

Diseases of Fishes, Book 2B, Identification of Fish Pathogenic Bacteria. G. L. Bullock. 1971. Neptune, N.J.: T. F. H. Publications. 41 pp.

Diseases of Fishes. Book 4, Fish Immunology. D. P. Anderson. 1974. Neptune, N.J.: T. F. H. Publications. 239 pp.

Diseases of Fishes, Book 5, Environmental Stress and Fish Diseases. G. A. Wedemeyer, F. P. Meyer, and L. Smith. 1976. Neptune, N.J.: T. F. H. Publications. 192 pp.

Fish Pathology, 2nd ed. R. J. Roberts, ed. 1989. London: Saunders. 448 pp.

Guidelines for the Use of Fishes in Field Research. C. Hubbs, J. G. Nickum, and J. R. Hunter. 1987. Joint publication of the American Society of Ichthyologists and Herpetologists, the American Fisheries Society, and the American Institute of Fisheries Research Biologists. 12 pp.

Guidelines for the Use of Live Amphibians and Reptiles in Field Research. V. H. Hutchinson, ed. 1987. Joint publication of the American Society of Ichthyologists and Herpetologists, The Herpetologists' League, and the Society for the Study of Amphibians and Reptiles. 14 pp.

Information Resources for Reptiles, Amphibians, Fish, and Cephalopods Used in Biomedical Research. D. J. Berry, M. D. Kreger, J. L. Lyons-Carter. 1992. Beltsville, Md.: USDA National Library Animal Welfare Information Center. 87 pp.

Laboratory Anatomy of the Turtle. L. M. Ashley. 1955. Dubuque, Iowa: Wm. C. Brown. 48 pp.

Parasites of Freshwater Fishes: A Review of Their Treatment and Control. G. L. Hoffman and F. P. Meyer. 1974. Neptune, N.J.: T. F. H. Publications. 224 pp.

The Pathology of Fishes. W. E. Ribelin and G. Migaki, eds. 1975. Madison: University of Wisconsin. 1004 pp.

ANESTHESIA, PAIN, AND SURGERY

Anesthesiology: Selected Topics in Laboratory Animal Medicine. Vol. 5. S. H. Cramlet and E. F. Jones. 1976. Brooks Air Force Base, Tex.: U.S. Air Force School of Aerospace Medicine. 110 pp. (Available as Accession No. ADA 031463 from National Technical Information Service, U.S. Department of Commerce, Springfield, VA 22161).

Animal Pain. Perception and Alleviation. R. L. Kitchell, H. H. Erickson, E. Carstens, and L. E. Davis. 1983. Bethesda, Md.: American Physiological Society. 231 pp.

Animal Pain Scales and Public Policy. F. B. Orlans. 1990. ATLA. 18:41-50.

Animal Physiologic Surgery. 2nd ed. C. M. Lang, ed. 1982. New York: Springer-Verlag. 180 pp.

Basic Surgical Exercises Using Swine. M. M. Swindle. 1983. New York: Praeger. 254 pp.

Canine Surgery: A Text and Reference Work. 2nd ed. J. Archibald, ed. 1974. Wheaton, Ill.: American Veterinary Publications. 1172 pp. (Publisher is now located in Santa Barbara, Calif.).

Categories of Invasiveness in Animal Experiments. Canadian Council on Animal Care. 1993. Guide to the Care and Use of Experimental Animals. Vol. 1 (2nd ed.). Appendix SV-B, pp. 201-202.

Comparative Anesthesia in Laboratory Animals. E. V. Miller, M. Ben, and J. S. Cass, eds. 1969. Fed. Proc. 28:1369-1586 and Index.

Experimental Surgery in Farm Animals. R. W. Dougherty. 1981. Ames: Iowa State University Press. 146 pp.

Experimental Surgery: Including Surgical Physiology. 5th ed. J. Markowitz, J. Archibald and H. G. Downie. 1964. Baltimore: Williams & Wilkins. 659 pp.

Experimental and Surgical Technique in the Rat. H. B. Waynforth and P. A. Flecknell. 1992. New York: Academic Press. 400 pp.

Fundamental Techniques in Veterinary Surgery. 3rd ed. C. B. Knocked, A. R. Allen, D. J. Williams, and J. H. Johnson. 1987. Philadelphia: W. B. Saunders. 368 pp.

Guidelines on the recognition of pain, distress and discomfort in experimental animals and an hypothesis for assessment. D. B. Morton and P. H. M. Griffiths. 1985. Vet. Rec. 116:431-436.

Laboratory Animal Anesthesia: An Introduction for Research Workers and Technicians. P. A. Flecknell. 1987. San Diego: Academic Press. 156 pp.

Large Animal Anesthesia: Principles and Techniques. T. W. Riebold, D. O. Goble, and D. R. Geiser. 1982. Ames: Iowa State University Press. 162 pp.

Pain, Anesthesia, and Analgesia in Common Laboratory Animals Bibliography, January 1980-December 1986. F. P. Gluckstein. 1986. Bethesda, Md.: National Library of Medicine (Publication #86-17). 45 pp.

Pain, Anesthesia, and Analgesia in Common Laboratory Animals Bibliography, January 1987 - May 1988. F. P. Gluckstein. 1988. Bethesda, Md.: National Library of Medicine (Publication #88-6). 9 pp.

Recognition and Alleviation of Pain and Distress in Laboratory Animals. NRC (National Research Council). 1992. A report of the Institute of Laboratory Animal Resources Committee on Pain and Distress in Laboratory Animals. Washington, D.C.: National Academy Press. 137 pp.

The Relief of Pain in Laboratory Animals. P. A. Flecknell. 1984. Lab. Anim. 18:147-160.

Research Animal Anesthesia, Analgesia, and Surgery. 1994. A. C. Smith and M. M. Swindle. Greenbelt, Md.: Scientists Center for Animal Welfare.

Small Animal Anesthesia: Mosby's Fundamentals of Animal Health Technology. R. G. Warren, ed. 1982. St. Louis: C. V. Mosby. 376 pp.

Small Animal Anesthesia: Mosby's Fundamentals of Animal Health Technology. D. McKelvey and W. Hollingshead. 1994. St. Louis: C. V. Mosby. 350 pp.

Small Animal Surgery. An Atlas of Operative Techniques. W. E. Wingfield and C. A. Rawlings, eds. 1979. Philadelphia: W. B. Saunders. 228 pp.

Small Animal Surgical Nursing. 2nd ed. Mosby's Fundamentals of Animal Health Technology. D. L. Tracy, ed. 1994. St. Louis: C. V. Mosby. 375 pp.

Standards for AAHA Hospitals. American Animal Hospital Association. 1990. Denver: AAHA. 71 pp.

Surgery of the Digestive System in the Rat. R. Lambert. 1965. (Translated from the French by B. Julien). Springfield, Ill.: Charles C Thomas. 501 pp.

Surgical Procedures. Laboratory Animal Science Association. 1990. Pp. 3-15 in Guidelines on the Care of Laboratory Animals and Their Use for Scientific Purposes III. London: Universities Federation for Animal Welfare.

Textbook of Large Animal Surgery. 2nd ed. F. W. Oehme and J. E. Prier. 1987. Baltimore: Williams & Wilkins. 736 pp.

Textbook of Small Animal Surgery. 2nd ed. D. Slatter. 1993. Philadelphia: W. B. Saunders. 2 Volumes. 2496 pp.

Textbook of Veterinary Anesthesia. L. R. Soma, ed. 1971. Baltimore: Williams & Wilkins. 621 pp.

Veterinary Anesthesia. 2nd ed. W. V. Lumb and E. W. Jones. 1984. Philadelphia: Lea and Febiger. 693 pp.

ANIMAL MODELS AND RESOURCES

Animal Models in Dental Research. J. M. Navia. 1977. University: University of Alabama Press. 466 pp.

Animal Models of Disease Bibliography, January 1979-December 1990. C. P. Smith. 1991. Beltsville, Md.: U.S. Department of Agriculture, National Agricultural Library. 31 pp.

Animal Models of Disease. K. J. Clingerman. 1991. Beltsville, Md.: U.S. Department of Agriculture, National Agricultural Library. 31 pp.

Animal Models of Thrombosis and Hemorrhagic Diseases. ILAR (Institute of Laboratory Animal Resources) Committee on Animal Models for Thrombosis and Hemorrhagic Diseases. 1976. DHEW Pub. No. (NIH) 76-982. Washington, D.C.: U.S. Department of Health, Education and Welfare. (Available from the Institute of Laboratory Animal Resources, National Research Council, 2101 Constitution Avenue, N.W., Washington, D.C. 20418).

Animals for Medical Research: Models for the Study of Human Disease. B. M. Mitruka, H. M. Rawnsley, and D. V. Vadehra. 1976. New York: John Wiley and Sons. 591 pp.

Bibliography of Induced Animal Models of Human Disease. G. Hegreberg and C. Leathers, eds. 1981. Pullman: Washington State University. 304 pp. (Available from Students Book Corporation, N.E. 700 Thatuna Street, Pullman, WA 99163).

Bibliography of Naturally Occurring Animal Models of Human Disease. G. Hegreberg and C. Leathers, eds. 1981. Pullman: Washington State University. 146 pp. (Available from Students Book Corporation, N.E. 700 Thatuna Street, Pullman, WA 99163).

The Future of Animals, Cells, Models, and Systems in Research, Development, Education and Testing. ILAR (Institute of Laboratory Animal Resources). 1977. Proceedings of a symposium organized by an ILAR committee. Washington, D.C.: National Academy of Sciences. 341 pp.

International Index of Laboratory Animals, 6th ed. 1993. Giving the location and status of over 7,000 stocks of laboratory animals throughout the world. Michael F. W. Festing, PO Box 301 Leicester, LE1 7RE, UK. 238 pp.

Mammalian Models for Research on Aging. NRC (National Research Council). 1981. A report of the ILAR (Institute of Laboratory Animal Resources) Committee on Animal Models for Research on Aging. Washington, D.C.: National Academy Press. 587 pp.

Resources for Comparative Biomedical Research: A Directory of the DRR Animal Resources Program. Research Resources Information Center. 1991. Bethesda, Md.: U.S. Department of Health and Human Services, Public Health Service, National Institutes of Health.

Spontaneous Animal Models of Human Disease. E. J. Andrews, D. C. Ward, and N. H. Altman, eds. 1979. Vol. 1, 322 pp.; Vol. 2, 324 pp. New York: Academic Press.

BIOHAZARDS IN ANIMAL RESEARCH

Animal-Associated Human Infections. A. N. Weinberg and D. J. Weber. 1991. Infectious Disease Clinics of North America 5:1-181.

Biohazards and Zoonotic Problems of Primate Procurement, Quarantine and Research. M. L. Simmons, ed. 1975. Cancer Research Safety Monograph Series, Vol. 2. DHEW Pub. No. (NIH) 76-890. Washington, D.C.: U.S. Department of Health, Education, and Welfare. 137 pp.

Biological Safety Manual for Research Involving Oncogenic Viruses. National Cancer Institute. 1976. DHEW Pub. No. 76-1165. Washington, D.C.: U.S. Department of Health, Education, and Welfare.

Biosafety in Microbiological and Biomedical Laboratories. 3rd ed. Centers for Disease Control and

National Institutes of Health. 1993. DHHS Pub. No. (CDC) 93-8395. Washington, D.C.: U.S. Department of Health and Human Services. 177 pp.

Biosafety in the Laboratory: Prudent Practices for Handling and Disposal of Infectious Materials. Committee on Hazardous Biological Substances in the Laboratory, National Research Council. 1989. Washington, D.C.: National Academy Press. 244 pp.

Classification of Etiologic Agents on the Basis of Hazard. 4th ed. U.S. Public Health Service Ad Hoc Committee on the Safe Shipment and Handling of Etiologic Agents. 1974. Washington, D.C.: U.S. Department of Health, Education, and Welfare.

Code of Federal Regulations. 1984. Title 40; Part 260, Hazardous Waste Management System: General; Part 261, Identification and Listing of Hazardous Waste; Part 262, Standards Applicable to Generators of Hazardous Waste; Part 263, Standards Applicable to Transporters of Hazardous Waste; Part 264, Standards for Owners and Operators of Hazardous Waste Treatment, Storage, and Disposal Facilities; Part 265, Interim Status Standards for Owners and Operators of Hazardous Waste Treatment, Storage, and Disposal Facilities; and Part 270, EPA Administered Permit Programs: The Hazardous Waste Permit Program. Washington, D.C.: Office of Federal Register. (Part 260, updated April 1994; 261 and 270 updated August, 1994; 264 and 265 updated June, 1994; 262 and 263 updated 1993).

Design Criteria for Viral Oncology Research Facilities. National Cancer Institute. 1975. DHEW Pub. No. (NIH)76-891. Washington, D.C.: U.S. Department of Health, Education, and Welfare. 24 pp.

Diseases Transmitted From Animals to Man. 6th ed. W. T. Hubbert, W. F. McCulloch, and P. R. Schnurrenberger, eds. 1974. Springfield, Ill.: Charles C Thomas. 1206 pp.

Guidelines for Carcinogen Bioassay in Small Rodents. J. M. Sontag, N. P. Page, and U. Saffiotti. 1976. DHEW Pub. No. (NIH) 76-801. Washington, D.C.: U.S. Department of Health, Education, and Welfare. 65 pp.

Guidelines for Research Involving Recombinant DNA Molecules. National Institutes of Health. 1984. Fed. Regist. 49(227):46266-46291.

Guidelines on Sterilization and High-Level Disinfection Methods Effective Against Human Immunodeficiency Virus (HIV). 1988. Geneva: World Health Organization. 11 pp.

Industrial Biocides. K. R. Payne, ed. 1988. New York: Wiley. 118 pp.

Laboratory Safety for Arboviruses and Certain Other Viruses of Vertebrates. Subcommittee on Arbovirus Safety, American Committee on Arthropod-Borne Viruses. 1980. Am. J. Trop. Med. Hyg. 29:1359-1381.

Laboratory Safety Monograph: A Supplement to the NIH Guidelines for Recombinant DNA Research. National Institutes of Health. 1979. Washington, D.C.: U.S. Department of Health, Education, and Welfare. 227 pp.

National Cancer Institute Safety Standards for Research Involving Oncogenic Viruses. National Cancer Institute. 1974. DHEW Pub. No. (NIH) 78-790. Washington, D.C.: U.S. Department of Health, Education, and Welfare. 20 pp.

NIH Guidelines for the Laboratory Use of Chemical Carcinogens. National Institutes of Health. 1981. NIH Pub. No. 81-2385. Washington, D.C.: U.S. Department of Health and Human Services. 15 pp.

An Outline of the Zoonoses. P. R. Schnurrenberger and W. T. Hubert. 1981. Ames: Iowa State University Press. 158 pp.

Prudent Practices in the Laboratory: Handling and Disposal of Chemicals. National Research Council. 1995. A report of the Committee on the Study of Prudent Practices for Handling, Storage, and Disposal of Chemicals in Laboratories. Washington, D.C.: National Academy Press.

The Zoonoses: Infections Transmitted from Animals to Man. J. C. Bell, S. R. Palmer, and J. M. Payne. 1988. London: Edward Arnold. 241 pp.

Zoonosis Updates from the Journal of the American Veterinary Medical Association. 1990. Schaumburg, Ill.: American Veterinary Medical Association. 140 pp.

BIRDS

American Ornithologists' Union. 1988. Report of Committee on Use of Wild Birds in Research. AUK. 105(1, Suppl):1A-41A.

Laboratory Animal Management: Wild Birds. NRC (National Research Council). 1977. A report of the ILAR (Institute of Laboratory Animal Resources) Committee on Standards, Subcommittee on Birds. 1977. Washington, D.C.: National Academy of Sciences. 116 pp.

Physiology and Behavior of the Pigeon. M. Abs, ed. 1983. London: Academic Press. 360 pp.

The Pigeon. W. M. Levi. 1974 (reprinted 1981). Sumter, S.C.: Levi Publishing. 667 pp.

Pigeon Health and Disease. D. C. Tudor. 1991. Ames: Iowa State University Press. 244 pp.

CATS AND DOGS

The Beagle as an Experimental Dog. A. C. Andersen, ed. 1970. Ames: Iowa State University Press. 616 pp.

Canine Anatomy: A Systematic Study. D. R. Adams. 1986. Ames: Iowa State University Press. 513 pp.

The Canine as a Biomedical Research Model: Immunological, Hematological, and Oncological Aspects. M. Shifrine and F. D. Wilson, eds. 1980. Washington, D.C.: Technical Information Center, U.S. Department of Energy. 425 pp. (Available as report no. DOE/TIC-10191 from National Technical Information Service, U.S. Department of Commerce, Springfield, VA 22161).

Laboratory Animal Management: Cats. ILAR (Institute of Laboratory Animal Resources) Committee on Cats. 1978. ILAR News 21(3):C1-C20.

Laboratory Animal Management: Dogs. NRC (National Research Council). 1994. A report of the ILAR (Institute of Laboratory Animal Resources) Committee on Dogs. Washington, D.C.: National Academy Press. 138 pp.

Miller's Anatomy of the Dog, 3rd ed. H. E. Evans. 1993. Philadelphia: W. B. Saunders. 1233 pp.

Textbook of Veterinary Internal Medicine: Diseases of the Dog and Cat. 3rd ed. 2 Vol. S. J. Ettinger, ed. 1989. Philadelphia: W. B. Saunders. 2464 pp.

DESIGN AND CONSTRUCTION OF ANIMAL FACILITIES

Approaches to the Design and Development of Cost-Effective Laboratory Animal Facilities. 1993. Canadian Council on Animal Care (CCAC) proceedings. Ottawa, Ontario, Canada: CCAC. 273 pp.

Comfortable Quarters for Laboratory Animals. rev. ed. 1979. Animal Welfare Institute. Washington, D.C.: Animal Welfare Institute. 108 pp.

Control of the Animal House Environment. T. McSheely, ed. 1976. London: Laboratory Animals Ltd. 335 pp.

Design of Biomedical Research Facilities. D. G. Fox, ed. 1981. Cancer Research Safety Monograph Series, Vol. 4. NIH Pub. No. 81-2305. Washington, D.C.: U.S. Department of Health and Human Services. 206 pp.

Design and Optimization of Airflow Patterns. S. D. Reynolds and H. Hughes. 1994. Lab Animal 23(9):46-49.

Estimating heat produced by laboratory animals. N. R. Brewer. 1964. Heat. Piping Air Cond. 36:139-141.

Guidelines for Construction and Equipment of Hospitals and Medical Facilities, 2nd ed. 1987. American Institute of Architects Committee on Architecture for Health. Washington, D.C.: American Institute of Architects Press. 111 pp.

Guidelines for Laboratory Design: Health and Safety Considerations. L. J. DiBerardinis, J. S. Baum, M. W. First, G. T. Gatwood, E. F. Groden, and A. K. Seth. 1993. New York: John Wiley & Sons. 514 pp.
Handbook of Facilities Planning. Volume 2: Laboratory Animal Facilities. T. Ruys, ed. 1991. New York: Van Nostrand Reinhold. 422 pp.
Laboratory Animal Houses: A Guide to the Design and Planning of Animal Facilities. G. Clough and M. R. Gamble. 1976. LAC Manual Series No. 4. Carshalton, Surrey, U.K.: Laboratory Animals Centre, Medical Research Council. 44 pp.
Laboratory Animal Housing. NRC (National Research Council). 1978. A report of the ILAR (Institute of Laboratory Animal Resources) Committee on Laboratory Animal Housing. Washington, D.C.: National Academy of Sciences. 220 pp
Structures and Environment Handbook. 11th ed. rev. Midwest Plan Service. 1987. Ames: Midwest Plan Service, Iowa State University.
The Use of Computational Fluid Dynamics For Modeling Air Flow Design in a Kennel Facility. H. C. Hughes and S. Reynolds. 1995. Contemp. Topics 34:49-53.

ENRICHMENT

Environmental Enrichment Information Resources for Nonhuman Primates: 1987-1992. National Agricultural Library, National Library of Medicine, and Primate Information Center. 1992. Beltsville, Md.: National Agricultural Library. 105 pp.
The Experimental Animal in Biomedical Research. Volume II: Care, Husbandry, and Well-being, An Overview by Species. B. E. Rollin and M. L. Kesel, eds. Boca Raton, Fla.: CRC Press.
Guidelines for developing and managing an environmental enrichment program for nonhuman primates. M. A. Bloomsmith, L. Y. Brent, and S. J. Schapiro. 1991. Laboratory Animal Science 41:372-377.
Housing, Care and Psychological Well-Being of Captive and Laboratory Primates. E. F. Segal, ed. 1989. Park Ridge, N.J.: Noyes Publications. 544 pp.
Monkey behavior and laboratory issues. K. Bayne and M. Novak, eds. Laboratory Animal Science 41:306-359.
The need for responsive environments. H. Markowitz and S. Line. 1990. Pp. 153-172 in The Experimental Animal in Biomedical Research. Volume I: A Survey of Scientific and Ethical Issues for Investigators, B. E. Rollin and M. L. Kesel, eds. Boca Raton, Fla.: CRC Press.
NIH Nonhuman Primate Management Plan. Office of Animal Care and Use. 1991. Bethesda, Md.: NIH, DHHS.
Psychological Well-Being of Nonhuman Primates. NRC (National Research Council). 1996. A report of the ILAR (Institute of Laboratory Animal Resources) Committee on Well-being of Nonhuman Primates. Washington, D.C.: National Academy Press.
Research and development to enhance laboratory animal welfare. 1992. R. A. Whitney. J. Am. Vet. Med. Assoc. 200(5):663-666.
A review of environmental enrichment strategies for single caged nonhuman primates. K. Fajzi, V. Reinhardt, and M. D. Smith. 1989. Lab Animal 18:23-35.
Through the Looking Glass. Issues of Psychological Well-Being in Captive Nonhuman Primates. M. Novak and A. J. Petto, eds. 1991. Washington, D.C.: American Psychological Association.

ENVIRONMENTAL CONTAMINANTS

Effect of environmental factors on drug metabolism: Decreased half-life of antipyrine in workers exposed to chlorinated hydrocarbon insecticides. B. Kolmodin, D. L. Azarnoff, and F. Sjoqvist. 1969. Clin. Pharmacol. Ther. 10:638-642.

Effect of essential oils on drug metabolism. A. Jori, A. Bianchett, and P. E. Prestini. 1969. Biochem. Pharmacol. 18:2081-2085.

Effect of intensive occupational exposure to DDT on phenylbutazone and cortisol metabolism in human subjects. A. Poland, D. Smith, R. Kuntzman, M. Jacobson, and A. H. Conney. 1970. Clin. Pharmacol. Ther. 11:724-732.

Effect of red cedar chip bedding on hexobarbital and pentobarbital sleep time. H. C. Ferguson. 1966. J. Pharm. Sci. 55:1142-1143.

Environmental and genetic factors affecting laboratory animals: impact on biomedical research. Introduction. C. M. Lang and E. S. Vesell. 1976. Fed. Proc. 35:1123-1124.

Environmental and genetic factors affecting the response of laboratory animals to drugs. E. S. Vesell, C. M. Lang, W. J. White, G. T. Passananti, R. N. Hill, T. L. Clemens, D. K. Liu, and W. D. Johnson. Fed. Proc. 35:1125-1132.

Frozen Storage of Laboratory Animals. G. H. Zeilmaker, ed. 1981. Stuttgart: Gustav Fischer. 193 pp.

Further studies on the stimulation of hepatic microsomal drug metabolizing enzymes by DDT and its analogs. L. G. Hart and J. R. Fouts. 1965. Arch. Exp. Pathol. Pharmakol. 249:486-500.

Induction of drug-metabolizing enzymes in liver microsomes of mice and rats by softwood bedding. E. S. Vesell. 1967. Science 157:1057-1058.

Influence on pharmacological experiments of chemicals and other factors in diets of laboratory animals. P. M. Newberne. 1975. Fed. Proc. 34:209-218.

The provision of sterile bedding and nesting materials with their effect on breeding mice. G. Porter and W. Lane-Petter. 1965. J. Anim. Tech. Assoc. 16:5-8.

ETHICS

Animal Liberation. 2nd ed. P. Singer. 1990. New York: New York Review Book. Distributed by Random House. 320 pp.

Animal Rights and Human Obligations, 2nd ed.. 1989. T. Regan and P. Singer. Englewood Cliffs, N.J.: Prentice-Hall. 280 pp.

The Assessment and 'Weighing' of Costs. In Lives in the Balance: The Ethics of Using Animals in Biomedical Research. J. A. Smith and K. Boyd, eds. 1991. London: Oxford University Press.

Ethical Scores for Animal Procedures. D. Porter. 1992. Nature 356:101-102.

The Experimental Animal in Biomedical Research. Volume I: A Survey of Scientific and Ethical Issues for Investigators. B. E. Rollin and M. L. Kesel, eds. 1990. Boca Raton, Fla.: CRC Press.

The Frankenstein Syndrome: Ethical and Social Issues in the Genetic Engineering of Animals. B. E. Rollin. 1995. New York: Cambridge University Press. 241 pp.

In the Name of Science: Issues in Responsible Animal Experimentation. F. B. Orlans. 1993. New York and Oxford: Oxford University Press.

Of Mice, Models, and Men: A Critical Evaluation of Animal Research. A. N. Rowan. 1984. Albany: State University of New York Press. 323 pp.

EUTHANASIA

Animal Euthanasia Bibliography. C. P. Smith and J. Larson. 1990. Beltsville, Md.: U.S. Department of Agriculture, National Agricultural Library. 31 pp.

Report of the AVMA panel on euthanasia. American Veterinary Medical Association. 1993. J. Am. Vet. Med. Assoc. 202(2):229-249.

EXOTIC, WILD, AND ZOO ANIMALS

Acceptable Field Methods in Mammalogy: Preliminary guidelines approved by the American Society of Mammalogists. American Society of Mammalogists. 1987. J. Mammal. 68(4, Suppl): 1-18.

Diseases of Exotic Animals: Medical and Surgical Management. 1983. Philadelphia: W. B. Saunders. 1159 pp.

Fur, Laboratory, and Zoo Animals. C. M. Fraser, J. A. Bergeron, and S. E. Aiello. 1991. Pp. 976-1087, Part IV, in The Merck Veterinary Manual, 7th ed. Rahway, N.J.: Merck and Co.

Kirk's Current Veterinary Therapy. Vol. XI. Small Animal Practice. R. W. Kirk and J. D. Bonagura, eds. 1992. Philadelphia: W. B. Saunders. 1388 pp.

The Management of Wild Mammals in Captivity. L. S. Crandall. 1964. Chicago: University of Chicago Press. 761 pp.

Pathology of Zoo Animals. L. A. Griner. 1983. San Diego, Calif.: Zoological Society of San Diego. 608 pp.

Restraint and Handling of Wild and Domestic Animals. M. E. Fowler. 1978. Ames: Iowa State University Press. 332 pp.

Zoo and Wild Animal Medicine. M. E. Fowler, ed. 1993. Philadelphia: W. B. Saunders. 864 pp.

FARM ANIMALS

Behavior of Domestic Animals. B. L. Hart. 1985. New York: W. H. Freeman. 390 pp.

The Biology of the Pig. W. G. Pond and K. A. Houpt. 1978. Ithaca, N.Y.: Comstock Publishing. 371 pp.

The Calf: Management and Feeding. 5th ed. J. H. B. Roy. 1990. Boston: Butterworths.

Clinical Biochemistry of Domestic Animals. 4th ed. J. J. Kaneko, ed. 1989. New York: Academic Press. 932 pp.

Current Veterinary Therapy. Food Animal Practice. J. L. Howard, ed. 1981. Philadelphia: W. B. Saunders. 1233 pp.

Current Veterinary Therapy: Food Animal Practice Two. J. L. Howard, ed. 1986. Philadelphia: W. B. Saunders. 1008 pp.

Current Veterinary Therapy. Food Animal Practice Three. J. L. Howard, ed. 1992. Philadelphia: W. B. Saunders. 1002 pp.

Diseases of Poultry. 9th ed. B. W. Calnek et al., eds. 1991. Ames: Iowa State University Press. 944 pp.

Diseases of Sheep. R. Jensen. 1974. Philadelphia: Lea and Febiger. 389 pp.

Diseases of Swine. 7th ed. A. D. Leman et al., eds. 1992. Ames: Iowa State University Press. 1038 pp.

Domesticated Farm Animals in Medical Research. R. E. Doyle, S. Garb, L. E. Davis, D. K. Meyer, and F. W. Clayton. 1968. Ann. N.Y. Acad. Sci. 147:129-204.

Dukes' Physiology of Domestic Animals. 11th rev. ed. M. J. Swenson and W. O. Reece, eds. 1993. Ithaca, N.Y.: Comstock Publishing. 928 pp.

Essentials of Pig Anatomy. W. O. Sack. 1982. Ithaca, N.Y.: Veterinary Textbooks. 192 pp.

Farm Animal Housing and Welfare. D. H. Baxter, M. R. Baxter, J. A. C. MacCormack, et al., eds. 1983. Boston: Nijhoff. 343 pp.

Farm Animal Welfare, January 1979-April 1989. C. N. Bebee and J. Swanson, eds. 1989. Beltsville, Md.: U.S. Department of Agriculture, National Agricultural Library. 301 pp.

Farm Animals and the Environment. C. Phillips and D. Piggins, eds. 1992. Wallingford, state: CAB International. 430 pp.

Indicators Relevant to Farm Animal Welfare. D. Smidt, ed. 1983. Boston: Nijhoff. 251 pp.

Livestock behavior and the design of livestock handling facilities. T. Grandin. 1991. Pp. 96-125 in Handbook of Facilities Planning. Volume 2: Laboratory Animal Facilities, T. Ruys, ed. New York: Van Nostrand. 422 pp.

Management and Welfare of Farm Animals. 3rd ed. UFAW (Universities Federation for Animal Welfare). 1988. London: Bailliere Tindall. 260 pp.

Nematode Parasites of Domestic Animals and of Man. N. D. Levine. 1968. Minneapolis, Minn.: Burgess Publishing. 600 pp.

Pathology of Domestic Animals. 4th ed. K. V. Jubb et al., eds. 1992. Vol. 1, 780 pp.; Vol. 2, 653 pp. New York: Academic Press.

The Pig as a Laboratory Animal. L. E. Mount and D. L. Ingram. 1971. New York: Academic Press. 175 pp.

The Protection of Farm Animals, 1979-April 1989: Citations From AGRICOLA Concerning Diseases and Other Environmental Considerations. C. N. Bebee, ed. 1989. Beltsville, Md.: U.S. Department of Agriculture, National Agricultural Library. 456 pp.

Reproduction in Farm Animals. E. S. E. Hafez. 1993. Philadelphia: Lea and Febiger. 500 pp.

Restraint of Domestic Animals. T. F. Sonsthagen. 1991. American Veterinary Publications.

Ruminants: Cattle, Sheep, and Goats. Guidelines for the Breeding, Care and Management of Laboratory Animals. NRC (National Research Council). 1974. A report of the ILAR (Institute of Laboratory Animal Resources) Committee on Standards, Subcommittee on Standards for Large (Domestic) Laboratory Animals. Washington, D.C.: National Academy of Sciences. 72 pp.

The Sheep as an Experimental Animal. J. F. Heckler. 1983. New York: Academic Press. 216 pp.

Swine as Models in Biomedical Research. M. M. Swindle. 1992. Ames: Iowa State University Press.

Swine in Cardiovascular Research. Vol. 1 and 2. H. C. Stanton and H. J. Mersmann. 1986. Boca Raton, Fla.: CRC Press.

GENERAL REFERENCES

Biology Data Book. 2nd ed. P. L. Altman and D. S. Dittmer. Vol. 1, 1971, 606 pp.; Vol. 2, 1973, 1432 pp.; Vol. 3, 1974, 2123 pp. Bethesda, Md.: Federation of American Societies for Experimental Biology.

Disinfection, Sterilization, and Preservation, 4th ed. S. S. Block, ed. 1991. Philadelphia: Lea and Febiger. 1162 pp.

A Guided Tour of Veterinary Anatomy: Domestic Ungulates and Laboratory Mammals. J. E. Smallwood. 1992. Philadelphia: W. B. Saunders. 390 pp.

Health Benefits of Animal Research. W. I. Gay. 1985. Washington, D.C.: Foundation for Biomedical Research. 82 pp.

The Inevitable Bond: Examining scientist-animal interactions. H. Davis and D. Balfour, eds. 1992. Cambridge: Cambridge University Press.

Jones' Animal Nursing. 5th ed. D. R. Lane, ed. 1989. Oxford: Pergamon Press. 800 pp.

Laboratory Animals. A. A. Tuffery. 1995. London: John Wiley.

Science, Medicine, and Animals. National Research Council, Committee on the Use of Animals in Research. 1991. Washington, D.C.: National Academy Press. 30 pp.

Use of Laboratory Animals in Biomedical and Behavioral Research. National Research Council and Institute of Medicine, Committee on the Use of Laboratory Animals in Biomedical and Behavioral Research. 1988. Washington, D.C.: National Academy Press. 102 pp.

Virus Diseases in Laboratory and Captive Animals. G. Darai, ed. 1988. Boston: Nijhoff. 568 pp.

GENETICS AND NOMENCLATURE

Effective population size, genetic variation, and their use in population management. R. Lande and G. Barrowclough. 1987. Pp. 87-123 in Viable Populations for Conservation M. Soule, ed. Cambridge: Cambridge University Press.

Genetics and Probability in Animal Breeding Experiments. E. L. Green. 1981. New York: Oxford University Press. 271 pp.

Holders of Inbred and Mutant Mice in the United States. Including the Rules for Standardized Nomenclature of Inbred Strains, Gene Loci, and Biochemical Variants. D. D. Greenhouse, ed. 1984. ILAR News 27(2):1A-30A.

Inbred and Genetically Defined Strains of Laboratory Animals. P. L. Altman and D. D. Katz, eds. 1979. Part 1, Mouse and Rat, 418 pp.; Part 2, Hamster, Guinea Pig, Rabbit, and Chicken, 319 pp. Bethesda, Md.: Federation of American Societies for Experimental Biology.

International Standardized Nomenclature for Outbred Stocks of Laboratory Animals. Issued by the International Committee on Laboratory Animals. M. Festing, K. Kondo, R. Loosli, S. M. Poiley, and A. Spiegel. 1972. ICLA Bull. 30:4-17 (March 1972). (Available from the Institute of Laboratory Animal Resources, National Research Council, 2101 Constitution Avenue, N.W., Washington, D.C. 20418).

Research-Oriented Genetic Management of Nonhuman Primate Colonies. S. Williams-Blangero. 1993. Laboratory Animal Science 43:535-540.

Standardized Nomenclature for Transgenic Animals. 1992. ILAR (Institute of Laboratory Animal Resources) Committee on Transgenic Nomenclature. ILAR News 34(4):45-52.

LABORATORY ANIMAL CARE

Animals for Research: Principles of Breeding and Management. W. Lane-Petter, ed. 1963. New York: Academic Press. 531 pp.

The Biomedical Investigator's Handbook for Researchers Using Animal Models. Foundation for Biomedical Research. 1987. Washington, D.C.: Foundation for Biomedical Research. 86 pp.

The Experimental Animal in Biomedical Research. Volume II: Care, Husbandry, and Well-being, An Overview by Species. B. E. Rollin and M. L. Kesel, eds. Boca Raton, Fla.: CRC Press.

Guidelines for the Treatment of Animals in Behavioral Research and Teaching. Animal Behavior Society. 1995. Anim. Behav. 49:277-282.

Handbook of Laboratory Animal Science, 2 Vol. P. Svendson and J. Hau. 1994. Boca Raton, Fla.: CRC Press. 647 pp.

Laboratory Animal Medicine. J. G. Fox, B. J. Cohen, and F. M. Loew, eds. 1984. New York: Academic Press. 750 pp.

Laboratory Animals: An Annotated Bibliography of Informational Resources Covering Medicine-Science (Including Husbandry)-Technology. J. S. Cass, ed. 1971. New York: Hafner Publishing. 446 pp.

Laboratory Animals: An Introduction for New Experimenters. A. A. Tuffey, ed. 1987. Chichester: Wiley-Interscience. 270 pp.

Methods of Animal Experimentation. W. I. Gay, ed. Vol. 1, 1965, 382 pp.; Vol. 2, 1965, 608 pp.; Vol. 3, 1968, 469 pp.; Vol. 4, 1973, 384 pp.; Vol. 5, 1974, 400 pp.; Vol. 6, 1981, 365 pp. Vol. 7, Part A, 1986, 256 pp.; Vol. 7, Part B, 1986, 269 pp.; Vol. 7, Part C, 1989, 237 pp. New York: Academic Press.

Pheromones and Reproduction in Mammals. J. G. Vandenbergh, ed. 1983. New York: Academic Press. 298 pp.

Practical Animal Handling. R. S. Anderson and A. T. B. Edney, eds. 1991. Elmsford, N.Y.: Pergamon. 198 pp.

Practical Guide to Laboratory Animals. C. S. F. Williams. 1976. St. Louis: C. V. Mosby. 207 pp.

Recent Advances in Germ-free Research. S. Sasaki, A. Ozawa, and K. Hashimoto, eds. 1981. Tokyo: Tokai University Press. 776 pp.

Reproduction and Breeding Techniques for Laboratory Animals. E. S. E. Hafez, ed. 1970. Philadelphia: Lea and Febiger. 275 pp.

Restraint of Animals. 2nd ed. J. R. Leahy and P. Barrow. 1953. Ithaca, N.Y.: Cornell Campus Store. 269 pp.

The UFAW Handbook on the Care and Management of Laboratory Animals. 6th ed. UFAW (Universities Federation for Animal Welfare). 1987. New York: Churchill Livingstone.

LAWS, REGULATIONS, POLICIES

Animals and Their Legal Rights. Animal Welfare Institute. 1985. Washington, D.C.: Animal Welfare Institute.

State Laws Concerning the Use of Animals in Research. National Association for Biomedical Research. 1991. Washington, D.C.

NONHUMAN PRIMATES

Aging in Nonhuman Primates. D. M. Bowden, ed. 1979. New York: Van Nostrand Reinhold. 393 pp.

The Anatomy of the Rhesus Monkey (*Macaca mulatta*). C. G. Hartman and W. L. Strauss, Jr., eds. 1933. Baltimore: Williams & Wilkins. 383 pp. (Reprinted in 1970 by Hafner, New York).

An Atlas of Comparative Primate Hematology. H. J. Huser. 1970. New York: Academic Press. 405 pp.

Behavior and Pathology of Aging in Rhesus Monkeys. R. T. Davis and C. W. Leathrus, eds. 1985. New York: Alan R. Liss.

Breeding Simians for Developmental Biology. Laboratory Animal Handbooks 6. F. T. Perkins and P. N. O'Donoghue, eds. 1975. London: Laboratory Animals Ltd. 353 pp.

Captivity and Behavior—Primates in Breeding Colonies, Laboratories and Zoos. J. Erwin, T. L. Maple, and G. Mitchell, eds. 1979. New York: Van Nostrand Reinhold. 286 pp.

The Care and Management of Chimpanzees (*Pan troglodytes*) in Captive Environments. R. Fulk and C. Garland, eds. 1992. Asheboro: North Carolina Zoological Society.

Comparative Pathology in Monkeys. B. A. Lapin and L. A. Yakovleva. 1963. Springfield, Ill.: Charles C Thomas. 272 pp.

Diseases of Laboratory Primates. T. C. Ruch. 1959. Philadelphia: W. B. Saunders. 600 pp.

A Handbook of Living Primates: Morphology, Ecology, and Behaviour of Nonhuman Primates. J. R. Napier and P. H. Napier. 1967. London: Academic Press. 456 pp.

Handbook of Squirrel Monkey Research. L. A. Rosenblum and C. L. Coe, eds. 1985. New York: Plenum Press. 501 pp.

Laboratory Animal Management: Nonhuman Primates. ILAR (Institute of Laboratory Animal Resources) Committee on Nonhuman Primates, Subcommittee on Care and Use. 1980. ILAR News 23(2-3):P1-P44.

Laboratory Primate Handbook. R. A. Whitney, Jr., D. J. Johnson, and W. C. Cole. 1973. New York: Academic Press. 169 pp.

Living New World Monkeys (*Platyrrhini*). Vol. 1. P. Hershkovitz. 1977. Chicago: University of Chicago Press. 117 pp.

The Macaques: Studies in Ecology, Behavior, and Evolution. D. G. Lindburg. 1980. New York: Van Nostrand Reinhold. 384 pp.

Macaca mulatta. Management of a Laboratory Breeding Colony. D. A. Valerio, R. L. Miller, J. R. M. Innes, K. D. Courtney, A. J. Pallotta, and R. M. Guttmacher. 1969. New York: Academic Press. 140 pp.

Nonhuman Primates in Biomedical Research: Biology and Management. B. T. Bennett, C. R. Abee, and R. Henrickson, eds. 1995. New York: Academic Press. 428 pp.

Pathology of Simian Primates. R. N. T. W. Fiennes, ed. 1972. Part I, General Pathology; Part II, Infectious and Parasitic Diseases. Basel: S. Karger.

Primates: Comparative Anatomy and Taxonomy. Vol. 1-7. W. C. O. Hill, ed. 1953-1974. New York: Interscience Publishers.

The Primate Malarias. G. R. Coatney, W. E. Collins, McW. Warren, and P. G. Contacos. 1971. Washington, D.C.: U.S. Department of Health, Education, and Welfare. 366 pp.

Zoonoses of Primates. The Epidemiology and Ecology of Simian Diseases in Relation to Man. R. N. T. W. Fiennes. 1967. London: Weidenfeld and Nicolson. 190 pp.

NUTRITION

Control of Diets in Laboratory Animal Experimentation. ILAR (Institute of Laboratory Animal Resources) Committee on Laboratory Animal Diets. 1978. ILAR News 21(2):A1-A12.

Effect of Environment on Nutrient Requirements of Domestic Animals. National Research Council, . 1981. A report of the Board on Agriculture and Renewable Resources Subcommittee on Environmental Stress, Committee on Animal Nutrition. Washington, D.C.: National Academy Press. 152 pp.

Feeding and Nutrition of Nonhuman Primates. R. S. Harris, ed. 1970. New York: Academic Press. 310 pp.

Feeds and Feeding. 3rd ed. E. Cullison. 1982. Reston, Va.: Reston Publishing. 600 pp.

Nutrient Requirements of Beef Cattle. 6th rev. ed. NRC (National Research Council). 1984. Nutrient Requirements of Domestic Animals Series. A report of the Board on Agriculture Subcommittee on Beef Cattle Nutrition, Committee on Animal Nutrition. Washington, D.C.: National Academy Press. 90 pp.

Nutrient Requirements of Cats. rev. ed. NRC (National Research Council). 1986. Nutrient Requirements of Domestic Animals Series. A report of the Board on Agriculture and Renewable Resources Panel on Cat Nutrition, Subcommittee on Laboratory Animal Nutrition, Committee on Animal Nutrition. Washington, D.C.: National Academy of Sciences. 88 pp. (See also *Taurine Requirement of the Cat*).

Nutrient Requirements of Dairy Cattle. 6th rev. ed. NRC (National Research Council). 1989. Nutrient Requirements of Domestic Animals Series. A report of the Board on Agriculture and Renewable Resources Subcommittee on Dairy Cattle Nutrition, Committee on Animal Nutrition. Washington, D.C.: National Academy of Sciences. 168 pp.

Nutrient Requirements of Dogs. rev. ed. NRC (National Research Council). 1985. Nutrient Requirements of Domestic Animals Series. A report of the Board on Agriculture and Renewable Resources Subcommittee on Dog Nutrition, Committee on Animal Nutrition. Washington, D.C.: National Academy of Sciences. 88 pp.

Nutrient Requirements of Goats: Angora, Dairy, and Meat Goats in Temperate and Tropical Countries. NRC (National Research Council). 1981. Nutrient Requirements of Domestic Animals Series. A report of the Board on Agriculture and Renewable Resources Subcommittee on Goat Nutrition, Committee on Animal Nutrition. Washington, D.C.: National Academy Press. 84 pp.

Nutrient Requirements of Horses. 5th rev. ed. NRC (National Research Council). 1989. Nutrient Requirements of Domestic Animals Series. A report of the Board on Agriculture Subcommittee on Horse Nutrition, Committee on Animal Nutrition. Washington, D.C.: National Academy of Sciences. 112 pp.

Nutrient Requirements of Laboratory Animals. 4th rev. ed. NRC (National Research Council). 1995. Nutrient Requirements of Domestic Animals Series. A report of the Board on Agriculture, Subcommittee on Laboratory Animal Nutrition, Committee on Animal Nutrition. Washington, D.C.: National Academy Press. 173 pp.

Nutrient Requirements of Nonhuman Primates. NRC (National Research Council). 1978. Nutrient Requirements of Domestic Animals Series. A report of the Board on Agriculture and Renewable Resources Panel on Nonhuman Primate Nutrition, Subcommittee on Laboratory Animal

Nutrition, Committee on Animal Nutrition. Washington, D.C.: National Academy of Sciences. 83 pp.

Nutrient Requirements of Poultry. 9th rev. ed. NRC (National Research Council). 1994. Nutrient Requirements of Domestic Animals Series. A report of the Board on Agriculture Subcommittee on Poultry Nutrition, Committee on Animal Nutrition. Washington, D.C.: National Academy Press. 176 pp.

Nutrient Requirements of Rabbits. 2nd rev. ed. NRC (National Research Council). 1977. Nutrient Requirements of Domestic Animals Series. A report of the Board on Agriculture and Renewable Resources Subcommittee on Rabbit Nutrition, Committee on Animal Nutrition. Washington, D.C.: National Academy of Sciences. 30 pp.

Nutrient Requirements of Sheep. 6th rev. ed. NRC (National Research Council). 1985. Nutrient Requirements of Domestic Animals Series. A report of the Board on Agriculture Subcommittee on Sheep Nutrition, Committee on Animal Nutrition. Washington, D.C.: National Academy of Sciences. 112 pp.

Nutrient Requirements of Swine. 9th rev. ed. NRC (National Research Council). 1988. Nutrient Requirements of Domestic Animals Series. A report of the Board on Agriculture and Renewable Resources Subcommittee on Swine Nutrition, Committee on Animal Nutrition. Washington, D.C.: National Academy of Sciences. 104 pp.

Nutrition and Disease in Experimental Animals. W. D. Tavernor, ed. 1970. Proceedings of a Symposium organized by the British Small Animal Veterinary Association, the British Laboratory Animal Veterinary Association, and the Laboratory Animal Scientific Association. London: Bailliere, Tindall and Cassell. 165 pp.

Taurine Requirement of the Cat. NRC (National Research Council). 1981. A report of the Board on Agriculture and Renewable Resources Ad Hoc Panel on Taurine Requirement of the Cat, Committee on Animal Nutrition. Washington, D.C.: National Academy Press. 4 pp.

United States-Canadian Tables of Feed Composition. 3rd rev. ed. NRC (National Research Council). 1982. A report of the Board on Agriculture and Renewable Resources Subcommittee on Feed Composition, Committee on Animal Nutrition. Washington, D.C.: National Academy Press. 156 pp.

OTHER ANIMALS

The Care and Management of Cephalopods in the Laboratory. P. R. Boyle. 1991. Herts, U.K.: Universities Federation for Animal Welfare. 63 pp.

Handbook of Marine Mammals. S. H. Ridgway and R. J. Harrison, eds. 1991. New York: Academic Press. 4 Vol.

Laboratory Animal Management: Marine Invertebrates. NRC (National Research Council). 1981. A report of the ILAR (Institute of Laboratory Animal Resources) Committee on Marine Invertebrates. Washington, D.C.: National Academy Press. 382 pp.

The Marine Aquarium Reference: Systems and Invertebrates. M. A. Moe. 1989. Plantation, Fla.: Green Turtle Publications. 510 pp.

The Principal Diseases of Lower Vertebrates. H. Reichenbach-Klinke and E. Elkan. 1965. New York: Academic Press. 600 pp.

PARASITOLOGY

Parasites of Laboratory Animals. R. J. Flynn. 1973. Ames: Iowa State University Press. 884 pp.

Veterinary Clinical Parasitology. 6th ed. M. W. Sloss and R. L. Kemp. 1994. Ames: Iowa State University Press. 198 pp.

PATHOLOGY AND CLINICAL PATHOLOGY

Atlas of Experimental Toxicological Pathology. C. Gopinath, D. E. Prentice, and D. J. Lewis. 1987. Boston: MTP Press. 175 pp.

An Atlas of Laboratory Animal Haematology. J. H. Sanderson and C. E. Phillips. 1981. Oxford: Clarendon Press. 473 pp.

Blood: Atlas and Sourcebook of Hematology, 2nd ed. C. T. Kapff and J. H. Jandl. 1991. Boston: Little, Brown. 158 pp.

Clinical Chemistry of Laboratory Animals. W. F. Loeb and F. W. Quimby. 1988. New York: Pergamon Press.

Clinical Laboratory Animal Medicine: An Introduction. D. D. Holmes. 1984. Ames: Iowa State University Press. 138 pp.

Color Atlas of Comparative Veterinary Hematology. C. M. Hawkey and T. B. Dennett. 1989. Ames: Iowa State University Press. .

Color Atlas of Hematological Cytology, 3rd ed. G. F. J. Hayhoe and R. J. Flemans. 1992. St. Louis: Mosby Year Book. 384 pp.

Comparative Neuropathology. J. R. M. Innes and L. Z. Saunders, eds. 1962. New York: Academic Press. 839 pp.

Essentials of Veterinary Hematology. N. C. Jain. 1993. Philadelphia: Lea and Febiger. 417 pp.

Immunologic Defects in Laboratory Animals. M. E. Gershwin and B. Merchant, eds. 1981. Vol. 1, 380 pp.; Vol. 2, 402 pp. New York: Plenum.

An Introduction to Comparative Pathology: A Consideration of Some Reactions of Human and Animal Tissues to Injurious Agents. G. A. Gresham and A. R. Jennings. 1962. New York: Academic Press. 412 pp.

Laboratory Profiles of Small Animal Diseases. C. Sodikoff. 1981. Santa Barbara, Calif.: American Veterinary Publications. 215 pp.

Outline of Veterinary Clinical Pathology. 3rd ed. M. M. Benjamin. 1978. Ames: Iowa State University Press. 352 pp.

Pathology of Laboratory Animals. K. Benirschke, F. M. Garner, and T. C. Jones. 1978. Vol. 1, 1050 pp.; Vol. 2, 2171 pp. New York: Springer Verlag.

The Pathology of Laboratory Animals. W. E. Ribelin and J. R. McCoy, eds. 1965. Springfield, Ill.: Charles C Thomas. 436 pp.

The Problems of Laboratory Animal Disease. R. J. C. Harris, ed. 1962. New York: Academic Press. 265 pp.

Roentgen Techniques in Laboratory Animals. B. Felson. 1968. Philadelphia: W. B. Saunders. 245 pp.

Schalm's Veterinary Hematology. 4th ed. O. W. Schalm and N. C. Jain. 1986. Philadelphia: Lea and Febiger. 1221 pp.

Techniques of Veterinary Radiography, 5th ed. J. P. Morgan, ed. Ames: Iowa State University Press. 482 pp.

Veterinary Clinical Pathology. 4th ed. E. H. Coles. 1986. Philadelphia: W. B. Saunders. 486 pp.

Veterinary Pathology. 5th ed. T. C. Jones and R. D. Hunt. 1983. Philadelphia: Lea and Febiger. 1792 pp.

PHARMACOLOGY AND THERAPEUTICS

Drug Dosage in Laboratory Animals: A Handbook. R. E. Borchard, C. D. Barnes, L. G. Eltherington. 1989. West Caldwell, N.J.: Telford Press.

Handbook of Veterinary Drugs: A Compendium for Research and Clinical Use. I. S. Rossoff. 1975. New York: Springer Publishing. 752 pp.

Mosby's Fundamentals of Animal Health Technology: Principles of Pharmacology. R. Giovanni and R. G. Warren, eds. 1983. St. Louis: C. V. Mosby. 254 pp.

Veterinary Applied Pharmacology and Therapeutics, 5th ed. G. C. Brander, D. M. Pugh, and R. J. Bywater. 1991. London: Bailliere Tindall. 624 pp.

Veterinary Pharmacology and Therapeutics. 6th rev. ed. N. H. Booth and L. E. McDonald. 1988. Ames: Iowa State University Press. 1238 pp.

RODENTS AND RABBITS

Anatomy and Embryology of the Laboratory Rat. R. Hebel and M. W. Stromberg. 1986. Worthsee, state: BioMed. 271 pp.

Anatomy of the Guinea Pig. G. Cooper and A. L. Schiller. 1975. Cambridge, Mass.: Harvard University Press. 417 pp.

Anatomy of the Rat. E. C. Greene. Reprinted 1970. New York: Hafner. 370 pp.

Bensley's Practical Anatomy of the Rabbit. 8th ed. E. H. Craigie, ed. 1948. Philadelphia: Blakiston. 391 pp.

The Biology and Medicine of Rabbits and Rodents. J. E. Harkness and J. E. Wagner. 1989. Philadelphia: Lea and Febiger. 230 pp.

The Biology of the Guinea Pig. J. E. Wagner and P. J. Manning, eds. 1976. New York: Academic Press. 317 pp.

Biology of the House Mouse. Symposia of the Zoological Society of London. No. 47. R. J. Berry, ed. 1981. London: Academic Press. 715 pp.

The Biology of the Laboratory Rabbit. S. H. Weisbroth, R. E. Flatt, and A. Kraus, eds. 1974. New York: Academic Press. 496 pp.

The Brattleboro Rat. H. W. Sokol and H. Valtin, eds. 1982. Ann. N.Y. Acad. Sci. 394:1-828.

Common Lesions in Aged B6C3F (C57BL/6N × C3H/HeN)F and BALB/cStCrlC3H/Nctr Mice. Syllabus. Registry of Veterinary Pathology, Armed Forces Institute of Pathology. 1981. Washington, D.C.: Armed Forces Institute of Pathology. 44 pp.

Common Parasites of Laboratory Rodents and Lagomorphs. Laboratory Animal Handbook. D. Owen. 1972. London: Medical Research Council. 140 pp.

Complications of Viral and Mycoplasmal Infections in Rodents to Toxicology Research and Testing. T. E. Hamm, ed. 1986. Washington, D.C.: Hemisphere Publishing. 191 pp.

Definition, Nomenclature, and Conservation of Rat Strains. ILAR (Institute of Laboratory Animal Resources) Committee on Rat Nomenclature. 1992. ILAR News 34(4): S1-S24.

A Guide to Infectious Diseases of Guinea Pigs, Gerbils, Hamsters, and Rabbits. NRC (National Research Council). 1974. A report of the ILAR (Institute of Laboratory Animal Resources) Committee on Laboratory Animal Diseases. Washington, D.C.: National Academy of Sciences. 16 pp.

Guidelines for the Well-Being of Rodents in Research. H. N. Guttman, ed. 1990. Bethesda, Md.: Scientists Center for Animal Welfare. 105 pp.

The Hamster: Reproduction and Behavior. H. I. Siegel, ed. 1985. New York: Plenum Press. 440 pp.

Handbook on the Laboratory Mouse. C. G. Crispens, Jr. 1975. Springfield, Ill.: Charles C Thomas. 267 pp.

Histological Atlas of the Laboratory Mouse. W. D. Gude, G. E. Cosgrove, and G. P. Hirsch. 1982. New York: Plenum. 151 pp.

Infectious Diseases of Mice and Rats. NRC (National Research Council). 1991. A report of the ILAR (Institute of Laboratory Animal Resources) Committee on Infectious Diseases of Mice and Rats. Washington, D.C.: National Academy Press. 397 pp.

Laboratory Anatomy of the Rabbit. 2nd ed. C. A. McLaughlin and R. B. Chiasson. 1979. Dubuque, Iowa: Wm. C. Brown. 68 pp.

Laboratory Animal Management: Rodents. NRC (National Research Council). In press. A report of the ILAR (Institute of Laboratory Animal Resources) Committee on Rodents. Washington, D.C.: National Academy Press.

A Laboratory Guide to the Anatomy of the Rabbit. 2nd ed. E. H. Craigie. 1966. Toronto: University of Toronto Press. 115 pp.

Laboratory Hamsters. G. L. Van Hoosier and C. W. McPherson, eds. 1987. New York: Academic Press. 456 pp.

The Laboratory Mouse: Selection and Management. M. L. Simmons and J. O. Brick. 1970. Englewood Cliffs, N.J.: Prentice-Hall. 184 pp.

The Laboratory Rat. H. J. Baker, J. R. Lindsey, and S. H. Weisbroth, eds. Vol. I, Biology and Diseases, 1979, 435 pp.; Vol. II, Research Applications, 1980, 276 pp. New York: Academic Press.

The Mouse in Biomedical Research. H. L. Foster, J. D. Small, and J. G. Fox, eds. Vol. I, History, Genetics, and Wild Mice, 1981, 306 pp.; Vol. II, Disease, 1982, 449 pp.; Vol. III, Normative Biology, Immunology, and Husbandry, 1983, 447 pp.; Vol. IV, Experimental Biology and Oncology, 1982, 561 pp. New York: Academic Press.

The Nude Mouse in Experimental and Clinical Research. J. Fogh and B. C. Giovanella, eds. Vol. 1, 1978, 502 pp.; Vol. 2, 1982, 587 pp. New York: Academic Press.

Origins of Inbred Mice. H. C. Morse III, ed. 1979. New York: Academic Press. 719 pp.

Pathology of Aging Rats: A Morphological and Experimental Study of the Age Associated Lesions in Aging BN/Bl, WAG/Rij, and (WAG × BN)F Rats. J. D. Burek. 1978. Boca Raton, Fla.: CRC Press. 230 pp.

Pathology of Aging Syrian Hamsters. R. E. Schmidt, R. L. Eason, G. B. Hubbard, J. T. Young, and D. L. Eisenbrandt. 1983. Boca Raton, Fla.: CRC Press. 272 pp.

Pathology of Laboratory Mice and Rats. Biology Databook Editorial Board. 1985. Bethesda, Md.: Federation of American Societies for Experimental Biology. 488 pp.

Pathology of the Syrian Hamster. F. Homburger, ed. 1972. Progr. Exp. Tumor Res. 16:1-637.

Proceedings of the Third International Workshop on Nude Mice. N. D. Reed, ed. 1982. Vol. 1, Invited Lectures/Infection/Immunology, 330 pp.; Vol. 2, Oncology, 343 pp. New York: Gustav Fischer.

The Rabbit: A Model for the Principles of Mammalian Physiology and Surgery. H. N. Kaplan and E. H. Timmons. 1979. New York: Academic Press. 167 pp.

Research Techniques in the Rat. C. Petty. 1982. Springfield, Ill.: Charles C Thomas. 368 pp.

Rodents and Rabbits: Current Research Issues. S. M. Niemi, J. S. Venable, and J. N. Guttman, eds. 1994. Bethesda, Md.: Scientists Center for Animal Welfare. 81 pp.

Viral and Mycoplasmal Infections of Laboratory Rodents: Effects on Biomedical Research. P. N. Blatt. 1986. Orlando, Fla.: Academic Press. 844 pp.

SAMPLE SIZE AND EXPERIMENTAL DESIGN

Animal welfare and the statistical consultant. R. M. Engeman and S. A. Shumake. 1993. American Statistician 47(3):229-233.

Appropriate animal numbers in biomedical research in light of animal welfare considerations. M. D. Mann, D. A. Crouse, and E. D. Prentice. 1991. Laboratory Animal Science 41:6-14.

The Design and Analysis of Long-Term Animal Experiments. J. J. Gart, D. Krewski, P. N. Lee, et al. 1986. Lyon: International Agency for Research on Cancer. 219 pp.

Power and Sample Size Review. T. J. Prihoda, G. M. Barnwell, and H. S. Wigodsky. 1992. Proceedings of the 1992 Primary Care Research Methods and Statistics Conference. Contact: Dr. T. Prihoda, Department of Pathology, University of Texas Health Science Center, San Antonio, TX 78284.

SERIAL PUBLICATIONS

Advances in Veterinary Science. Vol. 1-12. 1953-1968. New York: Academic Press.

Advances in Veterinary Science and Comparative Medicine (annual, continuation of Advances in Veterinary Science). New York: Academic Press.

The Alternatives Report (bimonthly). North Grafton, Ma.: Center for Animals & Public Policy, Tufts University.

American Journal of Pathology (monthly). Baltimore: American Society for Investigative Pathology.

American Journal of Primatology (monthly). New York: Wiley-Liss.

American Journal of Veterinary Research (monthly). Schaumburg, Ill.: American Veterinary Medical Association.

Animal Models of Human Disease (A Handbook). Washington, D.C.: The Registry of Comparative Pathology, Armed Forces Institute of Pathology.

The Animal Policy Report: A Newsletter on Animal and Environmental Issues (quarterly). North Grafton, Mass.: Center for Animals & Public Policy, Tufts University.

Animal Technology (semiannual, formerly The Institute of Animal Technicians Journal). Cardiff, U.K.: The Institute of Animal Technicians.

Animal Welfare (quarterly). Potters Bar, Herts, U.K.: Universities Federation for Animal Welfare.

Animal Welfare Information Center Newsletter (quarterly). Beltsville, Md.: Animal Welfare Information Center.

Animal Welfare Institute Quarterly. Washington, D.C.: Animal Welfare Institute.

ANZCCART News (quarterly). Glen Osmond, Australia: Australian and New Zealand Council for the Care of Animals in Research and Teaching.

Canadian Association for Laboratory Animal Medicine Newsletter. Canadian Association for Laboratory Animal Medicine.

Canadian Association for Laboratory Animal Science Newsletter. Canadian Association for Laboratory Animal Science.

Comparative Immunology, Microbiology and Infectious Diseases: International Journal for Medical and Veterinary Researchers and Practitioners (quarterly). Exeter, U.K.: Elsevier Science.

Comparative Pathology Bulletin (quarterly). Washington, D.C.: Registry of Comparative Pathology, Armed Forces Institute of Pathology.

Contemporary Topics (bimonthly). Cordova, Tenn.: American Association for Laboratory Animal Science.

Current Primate References (monthly). Seattle: Washington Regional Primate Research Center, University of Washington.

Folia Primatologica, International Journal of Primatology (6-weekly). Basel: S. Karger.

Humane Innovations and Alternatives (periodical). Washington Grove, Md.: Psychologists for the Ethical Treatment of Animals.

ILAR Journal (quarterly). Washington, D.C.: Institute of Laboratory Animal Resources (ILAR), National Research Council.

International Zoo Yearbook (annual). London: Zoological Society of London.

The Johns Hopkins Center for Alternatives to Animal Testing Newsletter (3 isses per year). Baltimore: Center for Alternatives to Animal Testing.

Journal of Medical Primatology (bimonthly). Copenhagen, Denmark: Munksgaard International Publishers.

Journal of Zoo and Wildlife Medicine (quarterly). Lawrence, Kans.: American Association of Zoo Veterinarians.

Lab Animal (11 issues per year). New York: Nature Publishers.

Laboratory Animal Science (bimonthly). Cordova, Tenn.: American Association for Laboratory Animal Science. Mailing address: 70 Timber Creek Dr., Cordova, Tenn. 38018.

Laboratory Animals (quarterly). Journal of the Laboratory Animal Science Association. London:

Laboratory Animals Ltd. Mailing address: The Registered Office, Laboratory Animals Ltd., 1 Wimpole Street, London W1M 8AE, United Kingdom.
Laboratory Primate Newsletter (quarterly). Providence, R.I.: Schrier Research Laboratory, Brown University.
Mouse News Letter (semiannual). Available to the western hemisphere and Japan from The Jackson Laboratory, Bar Harbor, ME 04609; available to other locations from Mrs. A. Wilcox, MRC Experimental Embryology and Teratology Unit, Woodmansterne Road, Carshalton, Surrey SM5 4EF, England.
Our Animal Wards. Washington, D.C.: Wards.
Primates: A Journal of Primatology (quarterly). Aichi, Japan: Japan Monkey Centre.
Rat News Letter (semiannual). Available from Dr. D. V. Cramer, ed., Department of Pathology, School of Medicine, University of Pittsburgh, Pittsburgh, PA 15261.
Resource. Ottawa, Ontario, Canada: Canadian Council on Animal Care.
SCAW Newsletter (quarterly). Bethesda, Md.: Scientists Center for Animal Welfare.
Zeitschrift fuer Versuchstierkunde, Journal of Experimental Animal Science (irregular, approximately 6 issues per year). Jena, Germany: Gustav Fischer Verlag.
Zoo Biology (bimonthly). New York: Wiley-Liss.
Zoological Society of London Symposia (annual). Oxford: Oxford Science.

TECHNICAL AND PROFESSIONAL EDUCATION

Clinical Textbook for Veterinary Technicians. 3rd ed. D. M. McCurnin. 1993. Philadelphia: W. B. Saunders. 816 pp.
Education and Training in the Care and Use of Laboratory Animals: A Guide for Developing Institutional Programs. National Research Council. 1991. A report of the Institute of Laboratory Animal Resources Committee on Educational Programs in Laboratory Animal Science. Washington, D.C.: National Academy Press. 152 pp.
The Education and Training of Laboratory Animal Technicians. S. Erichsen, W. J. I. van der Gulden, O. Hanninen, G. J. R. Hovell, L. Kallai, and M. Khemmani. 1976. Prepared for the International Committee on Laboratory Animals. Geneva: World Health Organization. 42 pp.
Educational Opportunities in Comparative Pathology-United States and Foreign Countries. Registry of Comparative Pathology, Armed Forces Institute of Pathology. 1992. Washington, D.C.: Universities Associated for Research and Education in Pathology. 51 pp.
Laboratory Animal Medicine: Guidelines for Education and Training. ILAR (Institute of Laboratory Animal Resources) Committee on Education. 1979. ILAR News 22(2):M1-M26.
Laboratory Animal Medicine and Science Audiotutorial Series. G. L. Van Hoosier, Jr., Coordinator. 1976-1979. Distributed by Health Sciences Learning Resources Center. University of Washington, Seattle.
Lesson Plans: Instructional Guide for Technician Training. 1990. AALAS (American Association for Laboratory Animal Science) Pub. No. 90-1. Joliet, Ill.: American Association for Laboratory Animal Science. 450 pp.
Training Manual Series, Vol. I., Assistant Laboratory Animal Technicians. AALAS (American Association for Laboratory Animal Science). 1989. AALAS Pub. No. 89-1. Joliet, Ill.: American Association for Laboratory Animal Science. 454 pp.
Training Manual Series, Vol. II., Laboratory Animal Technicians. AALAS (American Association for Laboratory Animal Science). 1990. AALAS Pub. No. 90-2. Joliet, Ill.: American Association for Laboratory Animal Science. 248 pp.
Training Manual Series, Vol. III, Laboratory Animal Technologist. AALAS (American Association for Laboratory Animal Science). 1991. AALAS Pub. No. 91-3. Joliet, Ill.: American Association for Laboratory Animal Science. 462 pp.

Syllabus of the Basic Principles of Laboratory Animal Science. Ad Hoc Committee on Education of the Canadian Council on Animal Care (CCAC). 1984. Ottawa, Ontario: Canadian Council on Animal Care. 46 pp. (Available from CCAC, 1105-151 Slater Street, Ottawa, Ontario K1P 5H3, Canada).

Syllabus for the Laboratory Animal Technologist. AALAS (American Association for Laboratory Animal Science). 1972. AALAS Pub. No. 72-2. Joliet, Ill.: American Association for Laboratory Animal Science. 462 pp.

WELFARE

Laboratory Animal Welfare Bibliography. W. T. Carlson, G. Schneider, J. Rogers, et al. 1988. Beltsville, Md.: U.S. Department of Agriculture, National Agricultural Library. 60 pp.

Laboratory Animal Welfare Bibliography. Scientists Center for Animal Welfare. 1988. Bethesda, Md.: Scientist Center for Animal Welfare. 60 pp.

Laboratory Animal Welfare. 1979-April 1989. C. N. Bebee, ed. 1989. Beltsville, Md.: U.S. Department of Agriculture, National Agricultural Library. 102 pp.

Laboratory Animal Welfare: Supplement 8. National Library of Medicine (NLM) Current Bibliographies in Medicine Series. Compiled by F. P. Gluckstein. 1992. CBM No. 92-2. Washington, D.C.: U.S. Department of Health and Human Services. 86 citations; 14 pp. (Available from Supt. of Docs., U.S. G.P.O.).

Scientific Perspective on Animal Welfare. W. J. Dodds and F. B. Orlans, eds. 1982. New York: Academic Press. 131 pp.

APPENDIX

B
Selected Organizations Related to Laboratory Animal Science

American Association for Accreditation of Laboratory Animal Care (AAALAC), 11300 Rockville Pike, Suite 1211, Rockville, MD 20852-3035 (phone: 301-231-5353; fax: 301-231-8282; e-mail: accredit@aaalac.org).

This nonprofit organization was formed in 1965 by leading U.S. scientific and educational organizations to promote high-quality animal care, use, and well-being and to enhance life-sciences research and education through a voluntary accreditation program. Any institution maintaining, using, importing, or breeding laboratory animals for scientific purposes is eligible to apply for AAALAC accreditation. The animal-care facilities of applicant institutions are visited and the program of animal care and use thoroughly evaluated by experts in laboratory animal science, who submit a detailed report to the Council on Accreditation. The council reviews applications and site-visit reports, using guidelines in the *Guide for the Care and Use of Laboratory Animals*, to determine whether full accreditation should be awarded. Accredited institutions are required to submit annual reports on the status of their animal facilities, and site revisits are conducted at intervals of 3 years or less. The Council on Accreditation reviews the annual and site-revisit reports to determine whether full accreditation should continue.

Fully accredited animal-care facilities receive a certificate of accreditation and are included on a list of such facilities published by the association. Many private biomedical organization strongly recommend that all grantees be supported by an AAALAC-accredited animal program. Full accreditation by AAALAC is accepted by the Office for Protection from Research Risks of the

National Institutes of Health as strong evidence that the animal facilities are in compliance with Public Health Service policy.

American Association for Laboratory Animal Science (AALAS), 70 Timber Creek Drive, Suite 5, Cordova, TN 38018 (phone: 901-754-8620; fax: 901-753-0046; e-mail: info@aalas.org; URL: http://www.aalas.org/).

AALAS is a professional, nonprofit organization of persons and institutions concerned with the production, care, and study of animals used in biomedical research. The organization provides a medium for the exchange of scientific information on all phases of laboratory animal care and use through its educational activities and certification. AALAS is dedicated to advancing and disseminating knowledge about the responsible care and use of laboratory animals for the benefit of human and animal life. AALAS publishes *Laboratory Animal Science* (bimonthly journal), *Contemporary Topics* (bimonthly journal), training manuals for laboratory animal technicians, an annual membership directory, a directory of certified technologists, and occasional pamphlets on special subjects. AALAS answers inquiries; conducts certification program for laboratory animal technicians; conducts annual scientific sessions at which original papers are presented, with seminars and workshops on laboratory animal science; distributes publications; lends film and slide sets; and makes referrals to other sources of information. Services are available to anyone.

American College of Laboratory Animal Medicine (ACLAM), Dr. Charles W. McPherson, Executive Director, 200 Summerwinds Drive, Cary, NC 27511 (phone: 919-859-5985; fax: 919-851-3126).

ACLAM is a specialty board recognized by the American Veterinary Medical Association (AVMA). It was founded in 1957 to encourage education, training, and research; to establish standards of training and experience for qualification; and to certify, by examination, qualified laboratory animal specialists as diplomates. To achieve these goals, the college seeks to interest veterinarians in furthering both training and qualifications in laboratory animal medicine.

The annual ACLAM Forum is a major continuing-education meeting. ACLAM also meets and sponsors programs in conjunction with the annual meetings of AVMA and the American Association for Laboratory Animal Science. It emphasizes and sponsors continuing-education programs; cosponsors symposia; cosponsors about 30 autotutorial programs on use, husbandry, and diseases of animals commonly used in research; and has produced 14 volumes on laboratory subjects, such as *The Laboratory* Rat and *The Mouse in Biomedical Research.*

American Humane Association (AHA), 236 Massachusetts Avenue, NE, Suite 203, Washington, D.C. 20002 (phone: 202-543-7780; fax: 202-546-3266).

AHA is a professional, nonprofit organization of organizations and individuals concerned with the exploitation, abuse, and neglect of children and animals. AHA was founded in 1877 and was the first national organization to protect children and animals.

AHA supports the 3 R's in biomedical research: refinement, reduction, and replacement where possible. AHA informs its members of issues in biomedical research through its magazine, *Advocate*, which is published quarterly.

American Society of Laboratory Animal Practitioners (ASLAP), Dr. Bradford S. Goodwin, Jr., Secretary-Treasurer, University of Texas, Medical School-CLAMC, 6431 Fannin Street, Room 1132, Houston, TX 77030-1501 (phone: 713-792-5127; fax: 713-794-4177).

ASLAP, founded in 1966, is open to any graduate of a veterinary college accredited or recognized by the American Veterinary Medical Association (AVMA) or Canadian Veterinary Medical Association (CVMA) who is engaged in laboratory animal practice and maintains membership in AVMA, CVMA, or any other national veterinary medical association recognized by AVMA. Its purpose is to disseminate ideas, experiences, and knowledge among veterinarians engaged in laboratory animal practice through education, training, and research at both predoctoral and postdoctoral levels. Two educational meetings are held annually, one each in conjunction with the annual meetings of AVMA and the American Association for Laboratory Animal Science.

American Society of Primatologists (ASP), Regional Primate Research Center, University of Washington, Seattle, WA 98195 (URL: http://www.asp.org).

The purposes of ASP are exclusively educational and scientific—specifically, to promote and encourage the discovery and exchange of information regarding primates, including all aspects of their anatomy, behavior, development, ecology, evolution, genetics, nutrition, physiology, reproduction, systematic, conservation, husbandry, and use in biomedical research. The ASP holds an annual meeting, sponsors the *American Journal of Primatology*, and publishes the ASP Bulletin quarterly. Any person engaged in scientific primatology or interested in supporting the goals of the society may apply for membership. Membership and information about the International Primatological Society can be obtained from ASP.

American Veterinary Medical Association (AVMA), 1931 North Meacham Road, Suite 100, Schaumburg, IL 60173-4360 (phone: 800-248-2862; fax: 708-925-1329; URL: http://www.avma.org/).

AVMA is the major national organization of veterinarians. Its objective is to advance the science and art of veterinary medicine, including its relationship to public health and agriculture. AVMA is the recognized accrediting agency for schools and colleges of veterinary medicine. It promotes specialization in veterinary medicine through the formal recognition of specialty-certifying organizations, including the American College of Laboratory Animal Medicine. The AVMA Committee on Animal Technician Activities and Training accredits 2-year programs in animal technology at institutions of higher learning throughout the United States. A list of accredited programs and a summary of individual state laws and regulations relative to veterinarians and animal technicians are available from AVMA.

Animal Welfare Information Center (AWIC), National Agricultural Library, 5th floor, Beltsville, MD 20705-2351 (phone: 301-504-6212; fax: 301-504-7125; e-mail: awic@nal.usda.gov; URL: http://netvet.wustl.edu/awic.htm or http://www.nalusda.gov).

AWIC, at the National Agricultural Library, was established by the 1985 amendments to the Animal Welfare Act. It provides information on employee training, improved methods of experimentation (including alternatives), and animal-care and animal-use topics through the production of bibliographies, workshops, resource guides, and *The Animal Welfare Information Center Newsletter.* AWIC services are geared toward those who must comply with the Animal Welfare Act, such as researchers, veterinarians, exhibitors, and dealers. Publications and additional information are available from AWIC.

Animal Welfare Institute (AWI), P.O. Box 3650, Washington, DC 20007 (phone: 202-337-2332; fax: 202-338-9478; e-mail: awi@igc.apc.org).

AWI is a nonprofit educational organization dedicated to reducing the pain and fear inflicted on animals by humans. Since its founding in 1951, AWI has promoted humane treatment of laboratory animals, emphasizing the importance of socialization, exercise, and environmental enhancement. The institute supports the "3 R's": replacement of experimental animals with alternatives, refinement to reduce animal pain and suffering, and reduction in the numbers of animals used. Educational material published by AWI includes the *AWI Quarterly, Comfortable Quarters for Laboratory Animals, Beyond the Laboratory Door, and Animals and Their Legal Rights* and is available free to scientific institutions and libraries and at cost to others. The institute welcomes correspondence and discussion with scientists, technicians, and IACUC members on improving the lives of laboratory animals.

Association of Primate Veterinarians (APV), Dr. Dan Dalgard, Secretary,

Corning Hazleton, 9200 Leesburg Turnpike, Vienna, VA 22162-1699 (phone: 703-893-5400 ext. 5390; fax: 703-759-6947).

APV is a nonprofit organization whose missions are to promote the dissemination of information related to the health, care, and welfare of nonhuman primates and to provide a mechanism by which primate veterinarians can speak collectively on matters regarding nonhuman primates. The organization developed after an initial workshop on the clinical care of nonhuman primates held in 1973 at the National Institutes of Health. Six years later, bylaws were adopted to formalize the missions and operation of the group. Members of APV are veterinarians who are concerned with the health, care, and welfare of nonhuman primates. The association meets annually, publishes a quarterly newsletter, and contributes to other scholarly and regulatory efforts and issues concerning nonhuman primates.

Australia and New Zealand Council for the Care of Animals in Research and Teaching (ANZCCART): ANZCCART Australia, The Executive Officer, PO Box 19, Glen Osmond, South Australia 5064, (phone: +61-8-303-7393; fax: +61-8-303-7113; e-mail: anzccart@waite.adelaide.edu.au; URL: http://www. adelaide.edu.au/ANZCCART/); ANZCCART New Zealand, The Executive Officer, C/- The Royal Society of New Zealand, PO Box 598 , Wellington, New Zealand (phone: +64-4-472 7421; fax: +64-4-473 1841; e-mail: anzccart@rsnz.govt.nz; URL: http://www.adelaide.edu.au/ANZCCART/).

ANZCCART was established in 1987 in response to concerns in both the scientific and the wider communities about the use of animals in research and teaching. ANZCCART is an independent body that has been developed to provide a national focus for these issues. Through its varied activities, ANZCCART seeks to promote effective communication and cooperation between all those concerned with the care and use of animals in research and teaching. ANZCCART's missions are to promote excellence in the care of animals used in research and teaching and thereby minimize their discomfort, to ensure that the outcomes of the scientific uses of animals are worthwhile, and to foster informed and responsible discussion and debate within the scientific and wider communities regarding the scientific uses of animals.

Canadian Association for Laboratory Animal Medicine/L'Association canadienne de la médecine des animaux de laboratoire (CALAM/ACMAL), Dr. Brenda Cross, Secretary-Treasurer, 102 Animal Resources Centre, 120 Maintenance Road, University of Saskatchewan, Saskatoon, Saskatchewan, Canada S7N 5C4.

CALAM/ACMAL is a national organization of veterinarians with an interest in laboratory animal medicine. The association's missions are to advise interested parties on all matters pertaining to laboratory animal medicine, to further the education of its members, and to promote ethics and professionalism in the field. The association is committed to the provision of appropriate veterinary care for all animals used in research, teaching, or testing. The association publishes a newsletter, *Interface*, four times a year.

Canadian Association for Laboratory Animal Science/L'association canadienne pour la technologie des animeaux laboratoire (CALAS/ACTAL), Dr. Donald McKay, Executive Secretary, CW401 Biological Science Building, Bioscience Animal Service, University of Alberta, Edmonton, Alberta, Canada T6G 2E9 (phone: 403-492-5193; fax: 403-492-7257; e-mail: dmckay @gpu.srv.ualberta.ca).

CALAS/ACTAL is composed of a multidisciplinary group of people and institutions concerned with the care and use of laboratory animals in research, teaching, and testing. The aims of the association are to advance the knowledge, skills, and status of those who care for and use laboratory animals; to improve the standards of animal care and research; and to provide a forum for the exchange and dissemination of knowledge regarding animal care and research. CALAS/ ACTAL maintains a Registry for Laboratory Animal Technicians, publishes a newsletter six times a year, and hosts an annual national convention.

Canadian Council on Animal Care (CCAC), Constitution Square, Tower II, 315-350 Albert, Ottawa, Ontario, Canada K1R 1B1 (phone: 613-238-4031; fax: 613-238-2837; e-mail: ccac@carleton.ca).

CCAC, founded in 1968 under the aegis of the Association of Universities and Colleges of Canada, became an independently incorporated, autonomous organization in 1982. Through its development of guidelines, assessment visits, and educational/consultation programs, the CCAC is the main advisory and re- view agency for the use of animals in Canadian science. Compliance with CCAC guidelines, published in two volumes, is a requirement for the receipt of grants or contracts. CCAC is currently funded by the Natural Sciences and Engineering Council of Canada, the Medical Research Council of Canada, and some federal departments.

Center for Alternatives to Animal Testing (CAAT), Johns Hopkins Univer- sity, 111 Market Place, Suite 840, Baltimore, MD 21202-6709 (phone: 410-223- 1693; fax: 410-223-1603; e-mail: caat@jhuhyg.sph.jhu.edu; URL: http:// infonet.welch.jhu.edu/~caat/).

CAAT was founded in 1981 to develop alternatives to the use of whole animals for product development and safety testing. Although CAAT's mission focuses primarily on the development of alternatives for testing, the center also works with organizations seeking to implement the 3 R's in research and education. These organizations are throughout the world, primarily in North America, Europe, Australia, and Japan.

CAAT is an academic research center based in the School of Hygiene and Public Health at Johns Hopkins University in Baltimore, whose programs encompass laboratory research, education/information, and validation of alternative methods.

CAAT's primary outreach to scientific and lay audiences its newsletter, which is published three times a year. A newsletter for middle-school students, *CAATALYST*, is published three times a year.

Center for Animals and Public Policy, Tufts University, School of Veterinary Medicine, 200 Westboro Road, N. Grafton, MA 01536 (phone: 508-839-7991; fax: 508-839-2953; e-mail: dpease@opal.tufts.edu).

The center is a unit of Tufts School of Veterinary Medicine that deals with all aspects of human-animal interactions. The center publishes two newsletters (*The Animal Policy Report*, quarterly; *The Alternatives Report*, bimonthly) and other reports and related items, including *The Animal Research Controversy*, a 200-page report that includes an appendix on the animal-protection movement. The center also has established an MS program in animals and public policy, a 1-year program directed at persons with a graduate degree or equivalent life experience.

Foundation for Biomedical Research (FBR), 818 Connecticut Avenue, NW, Suite 303, Washington, DC 20006 (phone: 202-457-0654; fax 202-457-0659; e-mail: nabr-fbr@access.digex.net; URL: http://www.fiesta.com/fbr).

FBR is a nonprofit, educational organization dedicated to promoting public understanding and support of the ethical use of animals in medical research. The Foundation has a wide range of educational materials available for students as well as the general public, including brochures, booklets, videotapes, speaker's kits, posters, and is a source of information on education and training materials related to laboratory animal science. FRB hosts press events and assists members of the media in locating researchers to address issues regarding animal research.

The Humane Society of the United States (HSUS), 2100 L Street, NW, Washington, DC 20037 (phone: 202-452-1100; fax: 301-258-3082; e-mail: HSUSLAB@ix.netcom.com).

HSUS is the nation's largest animal-protection organization. The society is active on a wide variety of humane issues, including those affecting wildlife, companion animals, and animals in laboratories and on farms. HSUS publishes a quarterly magazine (*The HSUS News*), a newsletter (*The Animal Activist Alert*), and a variety of reports, brochures, and other advocacy materials. The society works actively on issues involving the use of animals in research, safety testing, and education. This work is spearheaded by the HSUS Animal Research Issues Section, with the aid of a Scientific Advisory Council. The aims of this research are to promote the 3 R's of replacement, reduction, and refinement; strong regulations and their enforcement; openness and accountability among research institutions; and an end to egregious mistreatment of animals. HSUS pursues these aims through educational, legislative, legal, and investigative means. Staff are available to give presentations and write articles on these topics.

Institute of Laboratory Animal Resources (ILAR), National Research Council, National Academy of Sciences, 2101 Constitution Avenue, NW, Washington, DC 20418 (phone: 202-334-2590; fax: 202-334-1687; e-mail: ILAR@nas.edu; *ILAR Journal* e-mail: ILARJ@nas.edu; URL: http://www2.nas.edu/ilarhome).

ILAR develops guidelines and disseminates information on the scientific, technologic, and ethical use of animals and related biologic resources in research, testing, and education. ILAR promotes high-quality, humane care of animals and the appropriate use of animals and alternatives. ILAR functions within the mission of the National Academy of Sciences as an adviser to the federal government, the biomedical research community, and the public. *ILAR Journal* is published quarterly and is distributed to scientists, biomedical administrators, medical libraries, and students.

International Council for Laboratory Animal Science (ICLAS), Dr. Steven Pakes, Secretary General, Division of Comparative Medicine, University of Texas Southwestern Medical Center, 5323 Harry Hines Boulevard, Dallas, TX (phone: 214-648-3340; fax: 214-648-2659; e-mail: spakes@mednet.swmed.edu).

ICLAS is an international nongovernment scientific organization that was founded in 1961 under the auspices of UNESCO and several scientific unions. The aims of ICLAS are to promote and coordinate the development of laboratory animal science throughout the world, to promote international collaboration in laboratory animal science, to promote the definition and monitoring of quality laboratory animals, to collect and disseminate information on laboratory animal science, and to promote the humane use of animals in research, testing, and teaching through recognition of ethical principles and scientific responsibilities.

ICLAS has programs addressing microbiologic and genetic monitoring and standardization, assisting developing countries in pursuing their objectives in improving the care and use of laboratory animals, and improving education and training in laboratory animal science. ICLAS accomplishes its goals through regional scientific meetings, an international scientific meeting held every 4 years, the dissemination of information, and expert consultation with those requesting assistance.

ICLAS membership is composed of national members, scientific union members, scientific members, and associate members. The Governing Board is responsible for implementing the general policy of ICLAS and is elected by the General Assembly every 4 years.

Laboratory Animal Management Association (LAMA), Mr. Paul Schwikert, Past-President. P.O. Box 1744, Silver Spring, MD 20915 (phone: 313-577-1418; fax: 313-577-5890).

LAMA is a nonprofit educational organization. Membership includes individuals and institutions involved in laboratory animal management, medicine, and science. The mission of the association, founded in 1984, is to "enhance the quality of management and care of laboratory animals throughout the world." The objectives of LAMA include promoting the dissemination of ideas, experiences, and knowledge in the management of laboratory animals, encouraging continued education, acting as a spokesperson for the field of laboratory animal management, and assisting in the training of managers. The organization conducts a midyear forum on management issues and topics of interest to the general membership and an annual meeting in conjunction with the American Association of Laboratory Animals Science national meeting. *LAMA Review* is a quarterly journal on management issues published by the organization, and *LAMA Lines* is a bimonthly newsletter on topics of general interest to the membership.

Massachusetts Society for the Prevention of Cruelty to Animals/American Humane Education Society (MSPCA/AHES), 350 South Huntington Avenue, Boston, MA 02130 (phone: 617-522-7400; fax: 617-522-4885).

The Center for Laboratory Animal Welfare at MSPCA/AHES was formed in 1992 to bring thoughtful analysis to the complex issues surrounding the use of animals in research, testing, and education. Its work involves researching issues related to the welfare of laboratory animals, creating educational materials, and developing programs on issues of interest to the public.

Founded in 1868, MSPCA/AHES is one of the largest animal-protection organizations in the world. It operates three animal hospitals, seven animal shelters, and a statewide law-enforcement program in Massachusetts. It is widely recognized for national leadership in humane education, publications, legislative issues, and veterinary medicine.

National Association for Biomedical Research (NABR), 818 Connecticut Avenue, NW, Suite 303, Washington, DC 20006 (phone: 202-857-0540; fax 202-659-1902; e-mail: nabr-fbr@access.digex.net; URL: http://www.fiesta.com/nabr).

NABR is a nonprofit organization of 350 institutional members from both academia and industry whose mission is to advocate public policy that recognizes the vital role of laboratory animals in research, education, and safety testing. NABR is a source of information concerning existing and proposed animal welfare legislation and regulations at the national, state, and local level.

Office for Protection from Research Risks (OPRR), National Institutes of Health, 6100 Executive Blvd., Suite 3B01, Rockville, MD 20892 (phone: 301-496-7163; fax: 301-402-2803).

The Division of Animal Welfare of OPRR fulfills responsibilities set forth in the Public Health Service (PHS) Act. These include developing and monitoring, as well as exercising compliance oversight relative to, the PHS Policy on Humane Care and Use of Laboratory Animals (Policy), which applies to animals involved in research conducted or supported by any component of PHS; establishing criteria for and negotiation of assurances of compliance with institutions engaged in PHS-conducted or PHS-supported research using animals; directing the development and implementation of educational and instructional programs with respect to the use of animals in research; and evaluating the effectiveness of PHS policies and programs for the humane care and use of laboratory animals.

Primate Information Center, Regional Primate Research Center SJ-50, University of Washington, Seattle, WA 98195 (phone: 206-543-4376; fax: 206-865-0305).

The Primate Information Center's goal is to provide bibliographic access to all scientific literature on nonhuman primates for the research and educational communities. Coverage spans all publication categories (articles, books, abstracts, technical reports, dissertations, book chapters, etc.) and many subjects (behavior, colony management, ecology, reproduction, field studies, disease models, veterinary science, pharmacology, physiology, evolution, taxonomy, genetics, zoogeography, etc.). A comprehensive computerized database is maintained and used to publish a variety of bibliographic products to fulfill this mission. The collection of materials on primate research is fairly comprehensive. However, the center is an indexing service and not a library, so materials generally do not circulate. It will make individually negotiated exceptions for items that researchers are not able to acquire otherwise.

Primate Supply Information Clearinghouse (PSIC), Cathy A. Johnson-Delany, Director, Regional Primate Research Center, SJ-50 University of Washington, Seattle, WA 98195 (phone: 206-543-5178; fax: 206-685-0305; e-mail: cathydj@bart.rprc.washington.edu).

The goal of PSIC is to provide communication between research institutions, zoologic parks, and domestic breeding colonies for the efficient sharing of non-human primates and their tissues, equipment, and services. PSIC also publishes *New Listings* and the *Annual Resource Guide.*

Purina Mills, Inc., 505 North 4th and D Street, Richmond, IN 47374.

Purina Mills, Inc. offers a correspondence course, called Laboratory Animal Care Course, for everyone working with small animals. The course includes the following six lessons: introduction to laboratory animals; management of laboratory animals; housing, equipment, and handling; disease and control; glossary; and housing supplements and miscellaneous.

Scientists Center for Animal Welfare (SCAW), 7833 Walker Drive, Suite 340, Greenbelt, MD 20770 (phone: 301-345-3500; fax: 301-345-3503).

SCAW is an independent organization supported by individuals and institutions involved in research with animals and concerned about maintaining the highest standards of humane care. SCAW publishes resource materials, organizes conferences, and supports a wide variety of educational activities.

Universities Federation for Animal Welfare (UFAW), 8 Hamilton Close, South Mimms, Potters Bar, Herts EN6 3QD, United Kingdom (phone: 44-707-58202; fax: 44-707-49279).

UFAW was founded in 1926 as the University of London Animal Welfare Society (ULAWS). Its work expanded, and in order to allow a wider membership, UFAW was formed in 1938 with ULAWS as its first branch. UFAW publishes the *UFAW Handbook on the Care and Management of Laboratory Animals* and other publications.

United States Department of Agriculture, Animal and Plant Health Inspection Service, Regulatory Enforcement of Animal Care (REAC), 4700 River Road, Unit 84, Riverdale, MD 20737-1234 (phone: 301-734-4981; fax: 301-734-4328; e-mail: sstith@aphis.usda.gov).

The missions of the Animal Care Program are to provide leadership in establishing acceptable standards of humane animal care and treatment and to monitor and achieve compliance through inspections and educational and cooperative efforts. Copies of the Animal Welfare Regulations and the Animal Welfare Act are available from REAC.

Wisconsin Regional Primate Research Center (WRPRC) Library, University of Wisconsin, 1220 Capitol Court, Madison, WI 53715-1299 (phone: 608-263-3512; fax: 608-263-4031; e-mail: library@primate.wisc.edu; URL: http://www.primate.wisc.edu/WRPRC).

The library supports research programs of WRPRC and aids in the dissemination of information about nonhuman primates to the scientific community. Books, periodicals, newsletters, and other documents in all languages related to primatology are included. Special collections include rare books and audiovisual materials.

C
Some Federal Laws Relevant to Animal Care and Use

ANIMAL WELFARE

The Animal Welfare Act of 1966 (P.L. 89-544)—as amended by the Animal Welfare Act of 1970 (P.L. 91-579); 1976 Amendments to the Animal Welfare Act (P.L. 94-279); the Food Security Act of 1985 (P.L 99-198), Subtitle F (Animal Welfare File Name: PL99198); and the Food and Agriculture Conservation and Trade Act of 1990 (P.L. 101-624), Section 2503, Protection of Pets (File Name: PL101624)—contains provisions to prevent the sale or use of animals that have been stolen, to prohibit animal-fighting ventures, and to ensure that animals used in research, for exhibition, or as pets receive humane care and treatment. The law provides for regulating the transport, purchase, sale, housing, care, handling, and treatment of such animals.

Regulatory authority under the Animal Welfare Act is vested in the secretary of the U.S. Department of Agriculture (USDA) and implemented by USDA's Animal and Plant Health Inspection Service (APHIS). Rules and regulations pertaining to implementation are published in the Code of Federal Regulations, Title 9 (Animals and Animal Products), Chapter 1, Subchapter A (Animal Welfare). Available from: Regulatory Enforcement and Animal Care, APHIS, USDA, Unit 85, 4700 River Road, Riverdale, MD 20737-1234. File Name 9CFR93.

ENDANGERED SPECIES

The Endangered Species Act of 1973 (P.L. 93-205; 87 Statute 884) became effective on December 28, 1973, supplanting the Endangered Species Conservation Act of 1969 (P.L. 91-135; 83 Statute 275). The new law seeks "to provide a means whereby the ecosystems upon which endangered species and threatened species depend may be conserved, to provide a program for the conservation of such endangered species and threatened species, and to take such steps as may be appropriate to achieve the purposes of the treaties and conservation of wild flora and fauna worldwide."

Regulatory authority under the Endangered Species Act is vested in the secretary of the U.S. Department of the Interior (USDI) and implemented by USDI's Fish and Wildlife Service. Implementing rules and regulations are published in the Code of Federal Regulations, Title 50 (Wildlife and Fisheries), Chapter 1 (U.S. Fish and Wildlife Service, Department of the Interior), Subchapter B, Part 17 (Endangered and Threatened Wildlife and Plants). Copies of the regulations, including a list of species currently considered endangered or threatened, can be obtained by writing to the Office of Endangered Species, U.S. Department of the Interior, Fish and Wildlife Service, Washington, DC 20240.

D

Public Health Service Policy and Government Principles Regarding the Care and Use of Animals

PUBLIC HEALTH SERVICE POLICY ON HUMANE CARE AND USE OF LABORATORY ANIMALS

The *Public Health Service (PHS) Policy on Humane Care and Use of Laboratory Animals* was updated in 1996. In the policy statement, the PHS endorses the *U.S. Government Principles for the Utilization and Care of Vertebrate Animals Used in Testing, Research, and Training* (reprinted below), which were developed by the Interagency Research Animal Committee. The PHS policy implements and supplements these principles. Information concerning the policy can be obtained from the Office for Protection from Research Risks, National Institutes of Health, 6100 Executive Boulevard, MSC 7507, Rockville, MD 20892-7507.

PRINCIPLES FOR THE CARE AND USE OF ANIMALS USED IN TESTING, RESEARCH, AND TRAINING

The principles below were prepared by the Interagency Research Animal Committee. This committee, which was established in 1983, serves as a focal point for federal agencies' discussions of issues involving all animal species needed for biomedical research and testing. The committee's principal concerns are the conservation, use, care, and welfare of research animals. Its responsibilities include information exchange, program coordination, and contributions to policy development.

U.S. Government Principles for the Utilization and Care of Vertebrate Animals Used in Testing, Research, and Training

The development of knowledge necessary for the improvement of the health and well-being of humans as well as other animals requires *in vivo* experimentation with a wide variety of animal species. Whenever U.S. Government agencies develop requirements for testing, research, or training procedures involving the use of vertebrate animals, the following principles shall be considered; and whenever these agencies actually perform or sponsor such procedures, the responsible Institutional Official shall ensure that these principles are adhered to:

I.   The transportation, care, and use of animals should be in accordance with the Animal Welfare Act (7 U.S.C. 2131 et seq.) and other applicable Federal laws, guidelines, and policies.[1]

II.   Procedures involving animals should be designed and performed with due consideration of their relevance to human or animal health, the advancement of knowledge, or the good of society.

III.   The animals selected for a procedure should be of an appropriate species and quality and the minimum number required to obtain valid results. Methods such as mathematical models, computer simulation, and *in vitro* biological systems should be considered.

IV.   Proper use of animals, including the avoidance or minimization of discomfort, distress, and pain when consistent with sound scientific practices, is imperative. Unless the contrary is established, investigators should consider that procedures that cause pain or distress in human beings may cause pain or distress in other animals.

V.   Procedures with animals that may cause more than momentary or slight pain or distress should be performed with appropriate sedation, analgesia, or anesthesia. Surgical or other painful procedures should not be performed on unanesthetized animals paralyzed by chemical agents.

VI.   Animals that would otherwise suffer severe or chronic pain or distress that cannot be relieved should be painlessly killed at the end of the procedure or, if appropriate, during the procedure.

VII.   The living conditions of animals should be appropriate for their species and contribute to their health and comfort. Normally, the housing, feeding, and care of all animals used for biomedical purposes must be directed by a veterinarian or other scientist trained and experienced in the proper care, handling, and use of the species being maintained or studied. In any case, veterinary care shall be provided as indicated.

---

[1]For guidance throughout these Principles, the reader is referred to the *Guide for the Care and Use of Laboratory Animals* prepared by the Institute of Laboratory Animals Resources, National Academy of Sciences.

VIII.   Investigators and other personnel shall be appropriately qualified and experienced for conducting procedures on living animals. Adequate arrangements shall be made for their in-service training, including the proper and humane care and use of laboratory animals.

IX.   Where exceptions are required in relation to the provisions of these Principles, the decisions should not rest with the investigators directly concerned but should be made, with due regard to Principle II, by an appropriate review group such as an institutional animal care and use committee. Such exceptions should not be made solely for the purposes of teaching or demonstration.

Index

A

Accidents and emergencies, 17, 18, 46, 62
Acclimation and adaptation, 28-29
    of newly acquired animals, 57, 58
    to outdoor housing, 25, 30
Acquisition of animals, 57
Activity and exercise, 37, 38
Agricultural research, 4-5
Airborne contaminants, 17, 22-23, 24, 33, 62-63
Airflow, *see* Ventilation and airflow
Air pressure, 17, 31, 76, 79
Albinism, 35
Alternatives to animal research, 1, 10, 117
    recommended readings, 82-83
American Association for Accreditation of
    Laboratory Animal Care (AAALAC),
    102-103
American Association for Laboratory Animal
    Science (AALAS), 13, 103
American College of Laboratory Animal
    Medicine (ACLAM), 103
American Humane Association (AHA), 103-104
American Humane Education Society (AHES),
    110
American Society of Heating, Refrigeration,
    and Air-Conditioning Engineers
    (ASHRAE), 32

American Society of Laboratory Animal
    Practitioners (ASLAP), 104
American Society of Primatologists (ASP), 104
American Veterinary Medical Association
    (AVMA), 65, 104-105
Amphibians, recommended readings, 83
Analgesia and analgesics, 12, 64-65
    recommended readings, 83-84
Anesthesia and anesthetics, 12, 63, 64-65
    recommended readings, 83-85
    recovery from, 63-64, 79
    waste gases, 17
Animal and Plant Health Inspection Service
    (APHIS), 112-113, 114
Animal care and use protocols, 8-11
Animal training, 11, 25
Animal Welfare Act, 113, 114
Animal Welfare Information Center (AWIC),
    13, 37, 105
Animal Welfare Institute (AWI), 105
Animal Welfare Regulations (AWRs), 2, 4, 8,
    9, 10, 113, 114, 117
    housing guidelines, 25, 26
    on personnel qualifications, 13, 61
    on transport, 57
    on veterinary care, 13
Antibiotics, 61
Anxiolytics, 65

119

Heating, ventilation, and air conditioning
(HVAC) systems, 33, 34, 75-76
Heat loads, 30, 31, 32, 33
Height, of enclosures, 25, 26-27
HEPA filters (high-efficiency particulate air
filters), 33-34, 76
Herpesvirus simiae, 18, 59
Herpesvirus tamarinus, 59
Holidays, care during, 46
Horses, space requirements, 31, 38
Housing, 23, 117
factors in planning, 21-22
safety design, 15-16
see also Cages and caging; Outdoor
housing; Primary enclosures; Secondary
enclosures; Space requirements; *and*
specific animals
Humane Society of the United States (HSUS),
108-109
Human interaction, 38
Humidity and moisture, 22, 23, 24, 29, 30, 34,
75
Husbandry and management practices, 2, 38-46
for outdoor housing, 24-25
recommended readings, 92-93
see also Bedding; Cleaning and sanitation;
Food and feeding; Records and
recordkeeping; Repair and maintenance

I

Identification of animals, 46, 57
Illuminating Engineering Society of North
America (IESNA), 35
Illumination, 34-35, 76
Immunization, 18
Immunocompromised animals, 15, 44
Inbreeding, 47, 48
Incineration, 45
Infectious-disease studies, 18
Inspections, 9
Institute of Laboratory Animal Resources
(ILAR), 2, 13, 109
Institutional animal care and use committees
(IACUCs), 2, 3, 4, 5, 8, 9-10, 11, 22
and housing design, 24, 26, 27
recommended readings, 82
and surgical procedures, 12, 61
Interagency Research Animal Committee, 116
International Air Transport Association (IATA)
Live Animal Regulations, 57

International Council for Laboratory Animal
Science (ICLAS), 109-110
Isolation, *see* Quarantine; Separation and
isolation

L

Laboratory Animal Management Association
(LAMA), 110
Laboratory personnel
medical evaluation, 17-18
qualifications and training, 13-14, 117
recommended readings, 100-101
see also Occupational health and safety
Laundering services, 15
Light, 34-35, 76
Litter boxes, 26
Lymphocytic choriomeningitis virus, 60

M

Macaques, 18, 59
Macroenvironment, 22
Major survival surgery, 11-12, 61
Massachusetts Society for Prevention of
Cruelty to Animals (MSPCA), 110
Medical examinations (personnel), 17-18
Metabolic processes, 22-23
Mice, 35, 58
diseases, 59, 60
housing requirements, 27, 32
Microenvironment, 22-23, 31
Modeling, *see* Computer modeling
Moisture, *see* Humidity and moisture
Monkeys, *see* Nonhuman primates
Mouse hepatitis virus, 59, 60
Mycoplasma hyopneumoniae, 59

N

National Association for Biomedical Research
(NABR), 111
National Institutes of Health (NIH), 16, 18, 111
National Research Council (NRC), 14, 16, 18
Natural environments, 4, 22, 25
Neuromuscular blocking agents, 65
Nocturnal animals, 35
Noise, 17, 36, 73, 77
Nomenclature, 48
recommended readings, 91-92

Records and recordkeeping
  clinical, 46-47
  genetic management, 47-48
  identification, 46
Recycled airflow, 33-34, 76
Regulations, 2-3, 8, 10, 57
  recommended readings, 93
Regulatory Enforcement of Animal Care
    (REAC), 112-113
Removals from experiments, 10, 11, 12
Repair and maintenance, 23, 34
Reproduction, *see* Breeding and reproduction
Reptiles, recommended readings, 83
Respiratory protection, 17
Resting areas, 23, 25, 26, 36
Restraint, 11
Risk assessment, 14
Rodents, 36, 40, 43, 58
  albinism and photoxicity, 35
  diseases of, 59, 60
  housing requirements, 24, 25, 26, 27, 32
  inbreeding, 47
  pathogen-free, 60
  recommended readings, 97-98
  surgery on, 63, 78
  toe-clipping, 46
Runs, 23, 24, 42-43

S

Safety, *see* Occupational health and safety
Sanitation, *see* Cleaning and sanitation
Scientists Center for Animal Welfare (SCAW),
    112
Secondary enclosures, 22, 29, 44
  doors and windows, 73-74
  ventilation and airflow, 22, 31-33
  see also Cages and caging
Security, 46, 73
Sedation and sedatives, 12, 65
Separation and isolation, 46, 58-59, 60, 72
Serial publications, 99-100
Sheep, 26, 30, 38
Shelf-life, of food, 39
Sheltered housing, 24
Shelves, 24, 36
Showers, 15, 17, 73
Sialodacryoadenitis virus, 59
Simian hemorrhagic fever, 59
Simian immunodeficiency virus, 59

Social groups and social interaction, 21-27, 37-38, 40
Solid-bottom flooring, 24, 43
Space requirements, 25-28, 30, 31, 32, 38
Stabilization, 57, 58
Sterilization
  of cages and equipment, 24, 44
  of food, 39
  of hazardous wastes, 45
  for surgery, 62
Storage facilities and containers, 72, 77
  for food and bedding, 39, 40, 41, 77
  for waste, 45, 72, 77
Subclinical infections, 59, 60
Surgical procedures, 11-12, 60-64, 117
  autoclaving, 62, 79
  facilities, 62-63, 78-79
  oversight, 56
  postsurgical care, 63-64, 79
  recommended readings, 83-85
Surveillance, 16, 18, 58, 59-60
Swine
  diseases, 59
  space requirements, 26, 30

T

Temperature, 22, 24, 28-34, 74, 75
Tetanus, 18
Tethering, 11
"3 R's" (replacement, reduction, and
    refinement), 108, 109
Timed lighting systems, 35, 76
Toe-clipping, 46
Training, *see* Animal training; Education and
    training of personnel
Transgenic animals, 47-48
Transport of animals, 57
Traps, 46
Treatment and therapeutic procedures, 60
Tuberculosis, 18, 57
Tunnels, 36

U

Universities Federation for Animal Welfare
    (UFAW), 112
Urine, 23, 43
U.S. Department of Agriculture (USDA), 57,
    112-113, 114

Related Publications

The following publications are available from the National Academy Press, 2101 Constitution Avenue, NW, Lockbox 285, Washington, DC 20055 (phone toll-free 1-800-624-6242 or call 202-334-3313 in the Washington metropolitan area). You may also order electronically via Internet at http://www.nap.edu. Additional related publications, including the quarterly *ILAR Journal* and *Principles and Guidelines for the Use of Animals in Precollege Education*, are available directly from the Institute of Laboratory Animal Resources (phone: 202-334-2590; fax: 202-334-1687; email: ILAR@nas.edu; URL: http://www2.nas.edu/ilarhome/).

Occupational Health and Safety in the Care and Use of Research Animals.
  Forthcoming
Psychological Well-Being of Nonhuman Primates. Forthcoming
Rodents: Laboratory Animal Management Series. Forthcoming
Nutrient Requirements of Laboratory Animals, Fourth Revised Edition.
  1995
Dogs: Laboratory Animal Management Series. 1994
Nutrient Requirements of Poultry, Ninth Revised Edition. 1994
Nutrient Requirements of Fish. 1993
Recognition and Alleviation of Pain and Distress in Laboratory Animals.
  1992
Education and Training in the Care and Use of Laboratory Animals: A
  Guide for Developing Institutional Programs. 1991
Infectious Diseases of Mice and Rats. 1991